Natural Attenuation Considerations and Case Studies

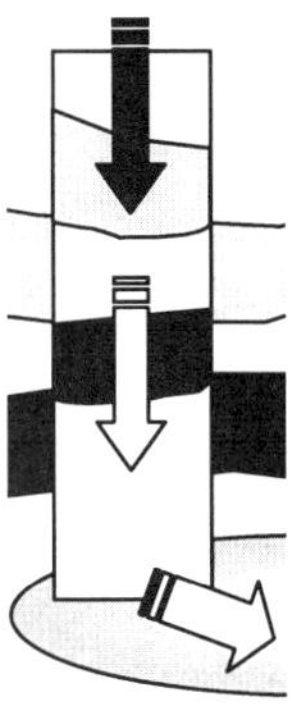

Remediation of Chlorinated and Recalcitrant Compounds

EDITORS
Godage B. Wickramanayake, Arun R. Gavaskar, and Mark E. Kelley
Battelle

The Second International Conference on Remediation of Chlorinated and Recalcitrant Compounds

Monterey, California, May 22–25, 2000

BATTELLE PRESS
Columbus • Richland

Library of Congress Cataloging-in-Publication Data

International Conference on Remediation of Chlorinated and Recalcitrant Compounds (2nd : 2000 : Monterey, Calif.)
Natural attenuation considerations and case studies : remediation of chlorinated and recalcitrant compounds (C2-3) / edited by Godage B. Wickramanayake, Arun R. Gavaskar, Mark E. Kelley.
p. cm.
"The Second International Conference on Remediation of Chlorinated and Recalcitrant Compounds, Monterey, California, May 22–25, 2000."
Includes bibliographical references and index.
ISBN 1-57477-097-7 (alk. paper)
1. Hazardous wastes--Natural attenuation--Congresses. 2. Hazardous wastes--Natural attenuation--Case studies--Congresses. 3. Organochlorine compounds--Biodegradation--Congresses. 4. Hazardous waste site remediation--Congresses. I. Wickramanayake, Godage B., 1953– II. Gavaskar, Arun R., 1962– III. Kelley, Mark E., 1960– IV. Title.

TD1066.O73 I58 2000d
628.5′2--dc21

00-034240

Printed in the United States of America

Battelle Press
505 King Avenue
Columbus, Ohio 43201-2693, USA
614-424-6393 or 1-800-451-3543
Fax: 1-614-424-3819
E-mail: press@battelle.org
Website: www.battelle.org/bookstore

For information on future environmental remediation meetings, contact:
Remediation Conferences
Battelle
505 King Avenue
Columbus, Ohio 43201-2693, USA
Fax: 614-424-3667
Website: www.battelle.org/conferences

CONTENTS

Natural Attenuation Case Studies I: Chloroethenes

Natural Attenuation Case Studies II: Other Organics and Metals

FOREWORD

Natural attenuation has become a viable remediation strategy for contaminated sites. What once was called a "do-nothing" approach has come to be regarded as a technically defensible, cost-effective remediation strategy and has been adopted at many sites contaminated with chlorinated and other recalcitrant compounds. *Natural Attenuation Considerations and Case Studies: Remediation of Chlorinated and Recalcitrant Compounds* brings together the results of recent research efforts and field applications in chapters that cover natural attenuation considerations, natural attenuation of chloroethenes, and natural attenuation of other organics and metals.

This is one of seven volumes resulting from the Second International Conference on Remediation of Chlorinated and Recalcitrant Compounds (May 22–25, 2000, Monterey, California). Like the first meeting in the series, which was held in May 1998, the 2000 conference focused on the more problematic contaminants—chlorinated solvents, pesticides/herbicides, PCBs/dioxins, MTBE, DNAPLs, and explosives residues—in all environmental media and on physical, chemical, biological, thermal, and combined technologies for dealing with these compounds. The conference was attended by approximately 1,450 environmental professionals involved in the application of environmental assessment and remediation technologies at private- and public-sector sites around the world.

A short paper was invited for each presentation accepted for the program. Each paper submitted was reviewed by a volume editor for general technical content. Because of the need to complete publication shortly after the Conference, no in-depth peer review, copy-editing, or detailed typesetting was performed for the majority of the papers in these volumes. Papers for 60% of the presentations given at the conference appear in the proceedings. Each section in this and the other six volumes corresponds to a technical session at the Conference. Most papers are printed as submitted by the authors, with resulting minor variations in word usage, spelling, abbreviations, the manner in which numbers and measurements are presented, and formatting.

We would like to thank the people responsible for the planning and conduct of the Conference and the production of the proceedings. Valuable input to our task of defining the scope of the technical program and delineating sessions was provided by a steering committee made up of several Battelle scientists and engineers – Bruce Alleman, Abraham Chen, James Gibbs, Neeraj Gupta, Mark Kelley, and Victor Magar. The committee members, along with technical reviewers from Battelle and many other organizations, reviewed more than 600 abstracts submitted for the Conference and determined the content of the individual sessions. Karl Nehring provided valuable advice on the development of the program schedule and the organization of the proceedings volumes. Carol Young, with assistance from Gina Melaragno, maintained program data, corresponded with speakers and authors, and compiled the final program and abstract books. Carol and Lori Helsel were responsible for the proceedings production effort, receiving assistance on specific aspects from Loretta Bahn, Tom Wilk, and Mark Hendershot. Lori, in particular, spent many hours examining papers for format and contacting authors as necessary to obtain revisions. Joe Sheldrick, the manager of Battelle Press, provided valuable production-planning advice; he and Gar Dingess designed the volume covers.

Battelle organizes and sponsors the Conference on Remediation of Chlorinated and Recalcitrant Compounds. Several organizations made financial contributions toward the 2000 Conference. The co-sponsors were EnviroMetal Technologies Inc. (ETI); Geomatrix Consultants, Inc.; the Naval Facilities Engineering Command (NAVFAC); Parsons Engineering Science, Inc.; and Regenesis.

As stated above, each article submitted for the proceedings was reviewed by a volume editor for basic technical content. As necessary, authors were asked to provide clarification and additional information. However, it would have been impossible to subject more than 300 papers to a rigorous peer review to verify the accuracy of all data and conclusions. Therefore, neither Battelle nor the Conference co-sponsors or supporting organizations can endorse the content of the materials published in these volumes, and their support for the Conference should not be construed as such an endorsement.

Godage B. Wickramanayake and Arun R. Gavaskar
Conference Chairs

NATIONAL RESEARCH COUNCIL GUIDANCE ON NATURAL ATTENUATION

Bruce E. Rittmann (Northwestern University, Evanston, IL)
Jacqueline A. MacDonald (RAND Corp., Washington, DC

ABSTRACT: With sponsorship from 5 government agencies and 5 private organizations, a National Research Council Committee was tasked to provide guidance on decision-making processes that ensure that natural attenuation is relied upon only when it truly protects human health and environmental quality. The Committee's report addresses four key aspects of natural attenuation as a remedy for groundwater contamination: (1) The essential role of community participation. (2) The scientific foundation for identifying when natural-attenuation processes are likely or unlikely to destroy or immobilize contaminants. (3) The type of evidence and level of interpretation that are needed to document that the natural-attenuation processes are protective. (4) A critical review of existing protocols for evaluation natural attenuation.

INTRODUCTION

Natural attenuation relies on unenhanced natural processes to protect human and environmental receptors from unacceptable exposures to groundwater contaminants. Many natural processes occur at contaminated sites. Some of those processes, particularly biodegradation and very strong immobilization, transform the contaminants to less harmful forms. Other natural processes dilute the contaminants or move them to other locations. Because natural attenuation is a legal remedy for waste-site cleanup, the presence of natural processes that may reduce contaminant concentration in the groundwater is not sufficient. Natural attenuation is an adequate remedy only when the natural processes are great enough to prevent harmful exposure for as long as the contaminant source persists.

Accepted as a remedial solution for BTEX (benzene, toluene, ethylbenzene, and xylene), natural attenuation is being proposed for chlorinated solvents, nitroaromatics, heavy metals, radionuclides, and other contaminants for which the scientific understanding and field experience are much less robust. Is the "band wagon" going too fast? Or, is the acceptance of natural attenuation being slowed unnecessarily due to apparent uncertainty?

In 1997, the National Research Council appointed the Committee on Intrinsic Remediation (the Committee) to provide guidance on when and how to select natural attenuation. The Committee's 15 members, listed in Table 1, included experts in the science and engineering disciplines key to natural-attenuation processes, the effects of these processes on contaminant fate, and the social/political factors important to the decision-making process. The findings of the Committee, reported in full in *Natural Attenuation for Groundwater Remediation* (NRC, 2000), are based on the expertise of the Committee members, detailed review of many documents and protocols on natural attenuation, and interviews of other experts and community leaders involved at contaminated sites. The report represents the consensus of the entire committee.

TABLE 1. NRC Committee on Intrinsic Remediation

Members	Affiliation
Bruce E. Rittmann, Chair	Northwestern University
Michael J. Barden	Geoscience Resources Ltd.
Barbara A. Bekins	U.S. Geological Survey
David Ellis	DuPont Specialty Chemicals
Stephen Lester	Center for Health, Environment, and Justice
Derek Lovley	University of Massachusetts
Richard G. Luthy	Carnegie-Mellon University*
Douglas M. Mackay	University of Waterloo
Eugene Madsen	Cornell University
Perry L. McCarty	Stanford University
Eileen Poeter	Colorado School of Mines
Robert Scofield	ENVIRON Corporation
Arthur W. Warrick	University of Arizona
John T Wilson	U.S. Environmental Protection Agency
John Zachara	Pacific Northwest National Laboratory
Others	
Jacqueline A. MacDonald	
Staff Officer	National Research Council**
James O. Leckie	
Board on Radioactive Waste Liaison	Stanford University

*Currently at Stanford University **Currently at RAND Corporation

This article summarizes the principal findings of the Committee's report (NRC, 2000). Those findings are divided into four subjects: community participation in decision-making; the scientific basis for natural attenuation; approaches for evaluating natural attenuation; and protocols for natural attenuation.

COMMUNITY PARTICIPATION IN DECISION-MAKING

When communities are aware of groundwater contamination, their representatives often are very skeptical of using natural attenuation as a formal remedy. Community members want the site cleaned up quickly, and they often object to remedies that leave significant quantities of contamination in place. Community leaders interviewed by the Committee expressed concern that natural attenuation allows responsible parties to save on cleanup costs, while shifting the risks to the community. Furthermore, they believed that observed reductions in contaminant concentration in the groundwater often come about because of dilution or transfer to another medium, where they continue to pose risks.

Community leaders indicated a willingness to accept natural attenuation when responsible parties and regulators can provide solid evidence that natural-attenuation processes are transforming the contaminants to harmless products. They generally are unwilling to accept dilution or other process that only move the contaminant from one place to another.

Public involvement in decision-making is especially important when natural attenuation is proposed as the remedy. On the one hand, community members must accept a large proportion of documented risks – health, economic, and social – for the long duration of a natural-attenuation remedy. On the other hand, decisions made without community participation may cause public outcry that can lead to delays or rejection of the natural-attenuation remedy. Although involving the affected public early in the decision-making process may slow the initial stages of remedy selection, early community involvement reduces delays and produces better decisions in the long run. The Report (NRC, 2000) reviews the results of past research and summarizes its interviews, both of which support the Committee's findings that public involvement is essential.

Three principles should guide community involvement: (1) involve the community early in the process; (2) provide the community with influence in the decision-making process; and (3) build an effective working relationship with the community. Public participation efforts at contaminated sites – particularly those for which natural attenuation is contemplated – need to be re-examined in light of the three principles.

To improve community participation when natural attenuation is being proposed, the Committee provides the recommendations in Table 2. These recommendations are tangible steps that can lead to achieving the three principles.

TABLE 2. Recommendations on community participation

Who	What
PRPs and Regulatory Agencies	Provide communities with clear evidence of which natural attenuation processes are responsible for loss of contamination
Regulatory Agencies	Change regulations to allow earlier community involvement
PRPs and Regulatory Agencies	Encourage community members to become involved in decision-making and oversight
Regulatory Agencies and PRPs	Ensure that community groups obtain independent technical advice
Regulatory Agencies and PRPs	Ensure that community groups can obtain all data concerning contamination and potential remedies

SCIENTIFIC BASIS FOR NATURAL ATTENTUATION

While natural attenuation has been documented as a remedy for situations in which the main groundwater contaminants are BTEX, it is not well established for other classes of contaminants. When favorable conditions are present, natural attenuation can be effective at sites with other contaminants, such as chlorinated solvents and metals. Its effectiveness will depend on site-specific conditions and the ability to document which natural-attenuation processes are acting. In some situations, special site conditions are necessary to make natural attenuation feasible. In other situations, the scientific understanding is so limited that it is difficult to document that natural attenuation is and will continue to protect public health and the environment.

The success of natural attenuation is contaminant and site-specific. Each contaminant is unique in terms of the reactions and environmental conditions that affect its fate. A major distinction is between organics and metals. While naturally occurring biodegradation can completely convert some organic contaminants to harmless products (e.g. H_2O, CO_2, and Cl^-), metals can only be transformed to forms that are more or less mobile. As for site conditions, the hydrogeological settings that favor natural attenuation for some contaminants are unfavorable for other contaminants. In addition, hydrogeological conditions can change with time and location.

Although decision-making about natural attenuation always is site-specific, the Committee provides general guidance on the likelihood that natural attenuation will succeed. This guidance is summarized in Tables 3 to 5. The judgments in these tables are based on our current level of understanding about the dominant processes that affect natural attenuation and the probability that sites will have the conditions needed to make the dominant processes effective. The Report (NRC, 2000) reviews the natural-attenuation processes and site settings that underlie these judgments. A high likelihood does not guarantee that natural attenuation will succeed at a particular site; a low likelihood does not mean that natural attenuation is impossible at a given site.

The Committee's conclusions about natural-attenuation potential are:

- Natural attenuation is well established as a remedy for only a few types of contaminants, mainly BTEX.
- Natural attenuation should never be considered a default or presumptive remedy. Its success must be documented at each site.
- Natural attenuation may need to continue for many years or decades.
- Natural attenuation process of some compounds can form hazardous byproducts, such as vinyl chloride (VC) from trichloroethene (TCE).
- Natural attenuation cannot destroy metals, but in some cases can immobilize them.
- The presence of mixtures can enhance or inhibit natural attenuation of one component.

TABLE 3. Summary of the likelihood of success for natural attenuation of nonhalogenated organic contaminants

Chemical or Chemical Class	Current Level of Understanding	Likelihood of Success
BTEX	High	High
Gasoline, Fuel Oil	Moderate	Moderate
Nonvolatile Aliphatics	Moderate	Low
Polycyclic Aromatics	Moderate	Low
Creosote	Moderate	Low
Low-MW Alcohols, Aldehydes, and Esters	High	High
MTBE	Moderate	Low
Nitroaromatics	Moderate	Low

Biotransformation is a dominant attenuation process in each case. Abiotic transformation or immobilization can play a key role in some cases. For full details, see NRC (2000).

TABLE 4. Summary of the likelihood of success for natural attenuation of halogenated organic contaminants

Chemical or Chemical Class	Current Level of Understanding	Likelihood of Success
TCE, PCE, CT	Moderate	Low
TCA	Moderate	Low
DCM	High	High
DCE, VC	Moderate	Low
Highly chlorinated aromatics	Moderate	Low
Lightly chlorinated aromatics	Moderate	Low
Monochlorobenzene	Moderate	Moderate

Biotransformation is a dominant attenuation process in each case. Immobilization can play a key role for highly chlorinated aromatics. For full details, see NRC (2000).

TABLE 5. Summary of the likelihood of success for natural attenuation of inorganic contaminants

Chemical or Chemical Class	Current Level of Understanding	Likelihood of Success
Metals		
Ni, Cu, Zn, Pb	Moderate	Low
Cd	Moderate	Low
Cr, Hg	Moderate	Low to Moderate
Nonmetals		
As, Se	Moderate	Low
Oxygenions		
NO_3^-	High	Moderate
ClO_4^-	Moderate	Low
Radionuclides		
^{60}CO, ^{137}Cs	Moderate	Moderate
^{3}H	High	Moderate

Immobilization, biotransformation, and/or radioactive decay can be dominant attenuation mechanisms. For full details, see NRC (2000).

APPROACHES FOR EVALUATING NATURAL ATTENUATION

Observing that contaminant concentrations decrease in groundwater samples is only one piece of evidence that natural attenuation is working. The decision to rely on natural attenuation and the confirmation that it continues to work requires establishing a cause-and-effect link to the underlying mechanisms. The Committee recommends that this link be made by using a site model and "footprints" of the underlying mechanisms. Footprints generally are concentration changes in reactants and products of the biogeochemical reactions that transform or immobilize the contaminants. Footprints occur because the natural-attenuation processes consume or produce other materials (e.g., oxygen, inorganic carbon, and chloride) in established stoichiometric ratios with the loss of the contaminant. Footprints are especially valuable because they can be detected in groundwater samples. Coupling the observed loss of contaminant to the measurement of several footprints is convincing evidence of which natural-attenuation processes are responsible for contaminant loss.

The basic steps for a site evaluation are:

1. Develop a site conceptual model that describes how the groundwater moves, where the contaminants are located, and which natural-attenuation processes are acting.
2. Analyze site measurements to confirm loss of contaminants, assess the formation of footprints, and determine if the natural-attenuation processes are sufficient to control contamination
3. Monitor the site to ensure that natural attenuation continues to work.

The level of effort required to carry out the three steps depends on the complexity of the site and the likelihood that the contaminant is controllable by natural attenuation (Tables 3 to 5). High complexity and low likelihood mean that the level of effort must increase. Table 6 summarizes the Committee's guidance about how these factors control level of effort.

TABLE 6. Summary of how site hydrogeology and likelihood of success affect the level of effort needed in site characterization and data interpretation

Site Characteristics			Likelihood of Success (Tables 3-5)		
Hydrogeology	Heterogeneity	Contaminant Concentration	High	Moderate	Low
Simple	Low	Low	1	2	2
Simple	Small Scale	Med. Or High	2	2	3
Transient	Large Scale	High	2	3	3

1=Low effort, 2=moderate effort, 3=high effort. For full details, see NRC (2000).

Level of effort includes the type and amount of samples taken, as well as the type of data interpretation. The Report (NRC, 2000) gives information and guidance on the number and location of samples, the number and type of footprints, and the kind of data analysis needed for each effort level. In particular, the Report describes how four levels of data analysis are appropriately employed:

1. Graphical and statistical analysis of trends for level 1 only.
2. Mass budgeting for level 2.
3. Simple transport modeling for levels 2 and 3.
4. Comprehensive transport modeling for level 3.

The Committee's recommendations for evaluating natural attenuation are:

- At every site when natural attenuation is being considered, responsible parties should use footprints of natural-attenuation processes to document which mechanisms are responsible for observed decreases in contaminant concentration.
- Responsible parties should prepare a conceptual model that describes groundwater flow, the contaminant source, the plume, and natural-attenuation processes significant to the site. The conceptual model should be tested and improved as data are gathered.
- Responsible parties should analyze field data at a level commensurate with the complexity of the site and the contaminant type.
- A long-term monitoring plan is needed for as long as natural attenuation is necessary to protect public health and the environment.

PROTOCOLS FOR NATURAL ATTENUATION

The Committee reviewed 14 natural-attenuation documents in detail. These documents, developed by a wide range of government and private organizations, provide different types of guidance concerning how to conduct studies on natural-attenuation potential. Most of them fall into the category of "protocols," which the Committee defines broadly to be an outline of a strategy and methodology to be followed for decision-making and implementation of natural attenuation (NRC, 2000).

With one exception, the available protocols address only fuel hydrocarbons or chlorinated solvents. The DOE protocol addresses metals and radionuclides, but the Committee thinks that the DOE protocol oversimplified the situation and must be substantially revised before it is used. No protocol addresses polycyclic aromatic hydrocarbons polychlorinated biphenyls, nitroaromatics, and a number of other classes (Tables 3 to 5).

The Committee compared the available protocols against characteristics of a comprehensive protocol summarized in Table 7. None of the existing protocols fulfills all of the characteristics in Table 7. To some degree, this situation reflects that a number of the documents were not designed to be comprehensive protocols. Significant and recurring "gaps" are in the characteristics of community involvement, institutional controls, contingency plans, sustainability, use qualifications, and peer review.

TABLE 7. **Summary of characteristics of a comprehensive protocol for natural attenuation**

Community Concerns

- Community involvement in decision-making.
- Institutional controls to restrict site use or access until cleanup goals are achieved.
- Contingency measures in place if natural attenuation fails in the future.

Scientific and Technical issues

- Cause-and-effect established from science understanding and field evidence.
- Site-assessment for hydrogeology and source.
- Sustainability established and documented.
- Peer-reviewed.

Implementation Issues

- User qualifications identified.
- Usability good.

For full details, see NRC (2000).

One special problem identified by the Committee relates to "scoring systems" that are intended to be used for initial screening. A numeric "score" is supposed to be used to establish whether or not a site is "eligible" to be considered for natural attenuation. Unfortunately, the scores frequently are used inappropriately to decide whether natural attenuation can be the remedy. A second special problem is that EPA does not officially endorse any protocols not developed by that agency. This situation creates ambiguity about what is acceptable for federal and state regulators.

The Committee recognizes that protocols can be very valuable to the decision-making process, but that current protocols fall short of being comprehensive. To improve the situation the Committee makes these recommendations.

- The EPA should lead an effort to develop national consensus guidelines for protocols on natural attenuation.
- The national consensus guidelines and all future protocols should be peer-reviewed, eliminate scoring systems in favor of the Committee's recommended evaluation approaches, and be easy to understand by non-technical audiences.
- EPA, other government agencies, and other organizations should provide additional training on natural attenuation concepts for regulators, site owners, remediation consultants, and community and environmental groups.

REFERENCE

National Research Council (NRC). 2000. *Natural Attenutation for Groundwater Remediation*. National Academy Press, Washington, DC.

REDOX CHARACTERIZATION FOR NATURAL ATTENUATION OF CHLORINATED ETHENES AND BTEX

Charlotte E. Riis and Anders G. Christensen (NIRAS, Alleroed, Denmark)
Rasmus Jakobsen (Dep. of Geology and Geotech. Engineering, Tech. University of DK)
Poul L. Bjerg (Dep. of Environmental Science and Engineering, Tech. University of DK)
Kim Dahlstrøm (Danish EPA, Copenhagen, Denmark)
Anne Mette G. Hansen (Municipality of Copenhagen, Copenhagen, Denmark)
Thomas Larsen (Hedeselskabet, Roskilde, Denmark)

ABSTRACT: Determination of the dominating redox processes in different ground water sections at a former industrial dry cleaning facility, contaminated with mineral spirits and chlorinated ethenes, was performed by evaluating changes in ground water geochemistry, levels of dissolved hydrogen and the in situ energy yields of the redox processes. Redox zonation at the site according to the H_2 levels correlates well with the zonation obtained by evaluating the redox sensitive species. Calculation of the energy yields of the relevant processes and comparison to literature threshold values and observed water chemistry resulted in generally the same zonation as the other two methods. The most detailed redox characterization at the site was attained from evaluation of the redox sensitive species. H_2 measurements are useful; however, the dependency of threshold values on temperature and electron acceptor concentrations may be considered when data are interpreted. The energetics of the redox processes are more suitable as verification and/or supplement for the redox characterization performed using the redox sensitive species than as an independent method, as long as precise threshold levels for the different processes are not known.

INTRODUCTION

One of the main problems regarding implementation of natural attenuation is to determine the extent of the degradation. Particularly for the chlorinated ethenes the redox conditions in an aquifer typically determine which organic compounds are degraded. Complete reductive dechlorination of PCE and TCE requires strongly reduced conditions. Therefore a detailed knowledge of the actual redox conditions is decisive for the prediction of the fate of the particular chlorinated compounds in the contaminant plume. Subsequently, it is important to examine which delineation methods give the most credible picture of the actual redox conditions in an aquifer.

Objective. The objective of this project has been to make a detailed redox characterization of the aquifer at a former industrial dry cleaning facility in Copenhagen, Denmark. Furthermore the objective has been to evaluate and quantify the degradation in the dissolved mineral spirit plume, to evaluate the degradation in the dissolved plumes of chlorinated ethenes and to evaluate the relation between the contaminant distribution, the redox conditions and the degradation at the site (Christensen and Riis, 2000). This paper presents a general description of three methods for characterizing the redox conditions in an aquifer along with a field-scale application of the methods.

Site Description. The geology at the site consists of glacial deposits dominated by fine sand present down to a minimum of 18 m below ground level (BGL), where the underlying bedrock (limestone) is found. The aquifer is unconfined and has an 8.5-m thick unsaturated zone.

Two plume types occur at the site, Figure 1d. One plume containing dissolved mineral spirits, represented by BTEX, and chlorinated ethenes, Figure 1a-c, and a second plume solely containing chlorinated ethenes, Figure 1b-c. As shown on Figure 1b there appears to be several (minimum 4) hot spots of chlorinated ethenes at the site.

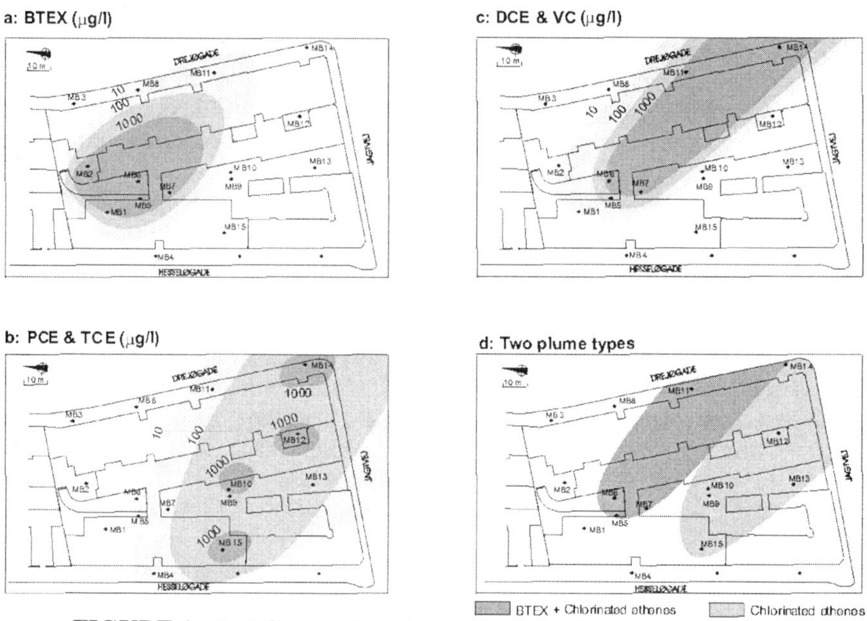

FIGURE 1. a)-c) Contaminant distribution. d) Two plume types.

METHODS

Redox sensitive species. Determination of the redox conditions in the ground water from the redox sensitive species was based on the consumption of electron acceptors (O_2, NO_3^-, SO_4^{2-}), the occurrence of intermediates (N_2O) and the production of reduced species (Fe^{2+}, Mn^{2+}, H_2S, CH_4) throughout the plume (e.g. Bjerg et al., 1995; Wiedemeier et al., 1996). The basis for this is that the organic material in the groundwater is degraded through aerobic respiration, NO_3^- reduction, Mn(IV) reduction, Fe(III) reduction, SO_4^{2-}reduction and methanogenesis. Similar approaches have previously been applied in various contaminated environments (Chapelle et al., 1995; Bjerg et al., 1995) Precipitation and dissolution of relevant minerals have been evaluated based on saturation indices calculated using the speciation program PHREEQC (Parkhurst et al., 1980).

Hydrogen level. The hydrogen distribution and concentration levels were evaluated and compared with the intervals proposed by Chapelle et al. (1995) and the ground water geochemistry at the site.

In situ energetics. The procedure for calculation of in situ energetics is described in Jakobsen et al., 1998. The energy yields outline the possibility that a given process can take place in a given aquifer and not whether the process will actually occur. By comparing calculated energy yields with the indications of processes seen in the water chemistry, an energy threshold can be estimated for each process, below which it is probable that the reaction actually occurs, because the microorganisms gain sufficient energy. The actual Gibbs energy of the redox processes is calculated according to the formulas given in Table 1.

TABLE 1. ΔG_r- expressions for the redox reactions.

Process	Gibbs energy of the reaction
Nitrate reduction	$\Delta G_r = \Delta G^0 + RT\ln([N_2]^{1/5}[H^+]^{-2/5}[NO_3^-]^{-2/5}[H_2]^{-1})$
Iron reduction	$\Delta G_r = \Delta G^0 + RT\ln([Fe^{2+}]^2[H^+]^{-4}[H_2]^{-1})$
Manganese reduction	$\Delta G_r = \Delta G^0 + RT\ln([Mn^{2+}][H^+]^{-2}[H_2]^{-1})$
Sulfate reduction	$\Delta G_r = \Delta G^0 + RT\ln([HS^-]^{1/4}[H^+]^{-1/4}[SO_4^{2-}]^{-1/4}[H_2]^{-1})$
Methanogenesis	$\Delta G_r = \Delta G^0 + RT\ln([CH_4]^{1/4}[H^+]^{-1/4}[HCO_3^-]^{-1/4}[H_2]^{-1})$

The calculated energy yields at the field site are compared to energy threshold values reported in literature (given in Table 2).

TABLE 2. Energy thresholds reported in literature.

Process	ΔG_r	Reference
Nitrate reduction	< -180 kJ/mole H_2 *	(Hoehler et al., 1998)
Iron reduction		
Ferrihydrite reduction	-70 to –113 kJ/mole H_2 ***	(Brandt, 1998)
Lepidocrocite reduction	-16 to –38 kJ/mole H_2 ***	(Brandt, 1998)
Sulfate reduction	-23 kJ/mole H_2 *	(Hoehler et al., 1998)
Methanogenesis	-20,4 kJ/mole H_2 *	(Hoehler et al., 1998)
Methanogenesis	-7 kJ/mole H_2 **	(Schulz & Conrad., 1996)

*Laboratory assay at in situ temperature and concentration of reactants and products. **Laboratory assay at 4°C and 20°C, respectively. *** Preliminary results from laboratory assays at 10°C with ground water sediment and synthetically produced ferrihydrite and lepidocrocite, respectively.

RESULTS AND DISCUSSION

Redox sensitive species. The unpolluted ground water (MB3 and MB4) is aerobic, characterized by relatively high concentrations of O_2, NO_3^- and SO_4^{2-}. CH_4 or H_2S is not detected, and the concentrations of Fe(II) and Mn(II) are low. The oxidation-reduction potential is positive, the pH is neutral and the alkalinity is high.

The ground water at the site has been divided in 3 redox levels (I, II and III), se Figure 2a, according to the water chemistry of the individual wells compared to the site specific criteria given in Table 3. The suggested criteria are set up based on the reactivity and transport properties of the specific redox sensitive species as well as from site specific concentration levels.

TABLE 3. Site specific criteria for redox level assessment based on redox sensitive species

	Aerobic respiration	Nitrate reduction	Iron reduction	Sulfate reduction	Methanogenesis
Oxygen	> 1 mg/l	<1 mg/l	<1 mg/l	<1 mg/l	<1 mg/l
Nitrate			< 30 mg/l	< 30 mg/l	< 30 mg/l
Dinitrogen oxide		YES			
Iron	< 1 mg/l	< 1 mg/l			
Manganese	< 0.5 mg/l	< 0.5 mg/l			
Sulfate					
Sulfide	0	0	0	YES	
Methane	< 1 mg/l	< 1 mg/l	< 1 mg/l	< 1 mg/l	>1 mg/l

Zone I covers the BTEX hot spot and the down gradient area (MB11), as shown on Figure 2a. There is no O_2, NO_3^- or N_2O. The Fe concentration is high in the BTEX hot spot and decreasing downgradient. The increase in Fe concentration compared with the background level indicates iron(III) reduction. The reason for the subsequent decrease in Fe concentration is that other processes remove the produced Fe(II). The Fe sink is hardly reoxidation, since there is no available O_2. The saturation indices for the Fe minerals indicate that a part of the Fe removal can be explained by precipitation of FeS and $FeCO_3$. The SO_4^{2-} concentrations in the BTEX hot spot are low compared to the background level. H_2S is found in most hot spot wells as a result of SO_4^{2-} reduction. The SO_4^{2-} concentrations increase in the downgradient part of zone I and the H_2S disappears rapidly. CH_4 has been observed throughout zone I, but not in the upgradient wells, indicating that the CH_4 observed in the BTEX hot spot is a result of on-going methanogenesis. As the CH_4 concentrations decrease downgradient in the plume, the CH_4 in the downgradient area is presumably transported from the BTEX hot spot being dispersed during transport. Based on the redox sensitive parameters the dominating redox processes in zone I are methanogenesis, SO_4^{2-} reduction and Fe(III) reduction.

Zone II is situated parallel to zone I, as shown in Figure 2a. The ground water in this zone is characterized by depletion of O_2. The lack of H_2S and CH_4 along with the high SO_4^{2-} concentrations excludes SO_4^{2-} reduction and methanogenesis. The concentrations of Fe and NO_3^- are low, but high levels of N_2O are found in this zone indicating ongoing NO_3^- reduction as the dominant redox process.

Zone III is found north and west of zones I and II (Figure 2a). O_2, NO_3^- and SO_4^{2-} are observed in the ground water in this zone. The Fe concentration is low, and H_2S, CH_4 and N_2O are not present. This indicates that zone III is aerobic.

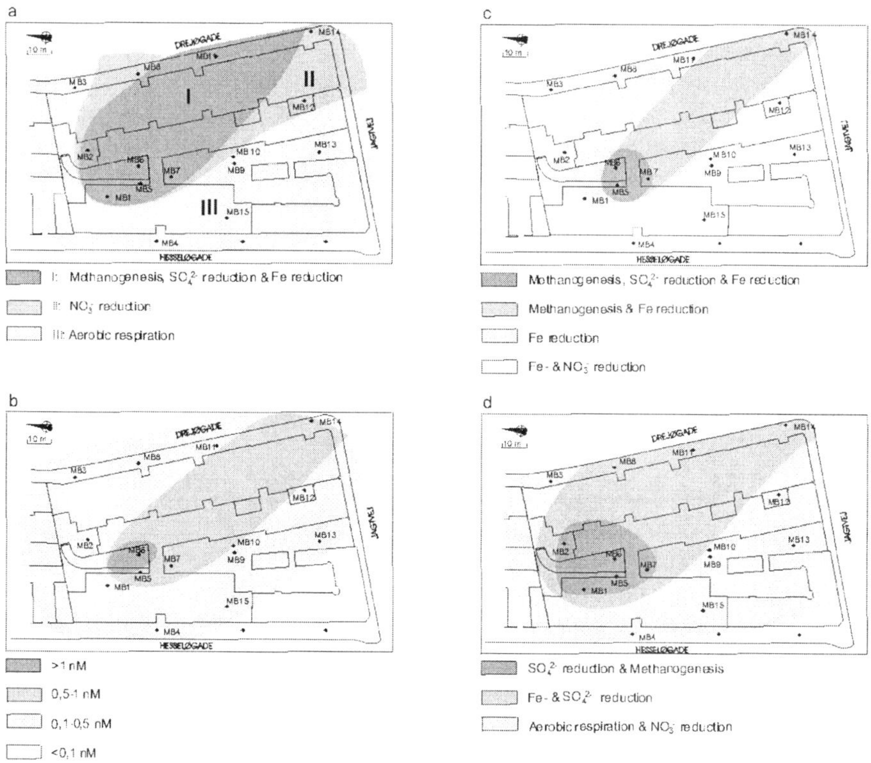

FIGURE 2a-d. Redox zonation at the field site according to: a) redox sensitive species, b) dissolved H_2, c) in situ energetics and d) suggested redox zonation.

Based on the redox sensitive parameters, several redox processes are occurring concomitantly, as observed by Bjerg et al. (1995), Ludvigsen et al. (1998), Jakobsen et al. (1998), Jakobsen and Postma (1999), and Hoehler (1998) as well.

Dissolved H_2. As shown in Figure 2b, the ground water at the site has been divided in 4 sections based on the distribution of H_2 concentrations. High H_2 concentrations are detected in the BTEX hot spot with decreasing concentrations in downgradient and outwards direction. If an evaluation of the observed H_2 intervals is made according to the criteria set up by Chapelle et al. (1995), SO_4^{2-} reduction is the dominating redox process in the BTEX hot spot. Fe(III) reduction dominates in the southwestern section of the site and NO_3^- reduction in the remaining part of the site. The criteria set up by Chapelle et al. (1995) apply to ground water with a temperature of 20^0C. It seems likely (Jakobsen et al., 1998; Hoehler et al., 1998) that the levels will be somewhat lower for ground water with at lower temperature. It is thus possible, that the actual processes in the particular sections are more reduced than assumed according to the criteria set up by Chapelle et al. (1995).

In situ energetics. The redox zonation derived from the calculated energy yields of the redox processes compared with the literature values in Table 2 is illustrated in Figure 2c. According to this, methanogenesis, SO_4^{2-} reduction and Fe(III) reduction are dominating in the BTEX hot spot, while methanogenesis and Fe(II) reduction are occurring in the rest of the plume. Fe(III) reduction is occurring in the section parallel to the plume, and Fe(III) reduction and NO_3^- reduction are taking place at the remaining part of the site. However, the delineation of the Fe(III) reduction will depend on the iron oxide mineralogy (Jakobsen et al., 1998; Postma and Jakobsen, 1996), since the less stable iron oxides are not necessarily present in the sediment. The calculated energy yields are compared with the observed water chemistry to estimate the site specific threshold values. The derived apparent threshold values at the field site are given in Table 4. They correspond well with threshold values reported in literature, see Table 2.

TABLE 4. Derived apparent energy thresholds, field site.

Process	Observed ΔG_r
Nitrate reduction	< -220 kJ/mole H_2[1]
Lepidocrocite reduction	-25 kJ/mole H_2[3]
Sulfate reduction	-25 kJ/mole H_2[1]
Methanogenesis	-20 kJ/mole H_2[2]

IMPACT ON FATE OF PCE AND TCE

Two plume types occur at the site, Figure 1d. One plume consisting of BTEX and chlorinated ethenes. In the hot spot of this plume, where the redox conditions are strongly reducing, a 99,5% transformation of PCE is observed. The major degradation products are cisDCE and VC (Figure 1c) indicating reductive dechlorination by a biological process. Three percent of the PCE is subject to complete reductive dechlorination to ethene. Within 60 meters downgradient of the hot spot, 96% of the BTEX is degraded. Adjacent to this plume is a second plume of chlorinated ethenes, Figure 1b-d. The redox conditions in this plume are aerobic/slightly reducing and the dechlorination is of limited extent. A 5% transformation of PCE is observed. None of the PCE is completely dechlorinated to ethene. The above mentioned percentages are valid within the observed distance (ca. 60 meters downgradient of the BTEX hot spot).

CONCLUSIONS

The suggested redox zonation based on the assessments according to the three methods outlined above is illustrated in Figure 2d. The most detailed characterization of the redox zones at the site was attained by the assessment from the redox sensitive species. Data from the present case indicate that zonation of the site according to the H_2 levels correlates well with the zonation obtained by evaluating the redox sensitive species. Calculation of the energy yields of the relevant processes and comparison to literature threshold values and observed water chemistry resulted in generally the same zonation as the other two methods.

The data show a close correlation between the contaminant distribution, the redox conditions and the extent of the reductive dechlorination. The most complete and extensive dechlorination is occurring in the strongly reduced section of the aquifer, while the dechlorination in the aerobic/slightly reduced section is limited.

ACKNOWLEDGEMENTS

We greatly recognize the funding provided by the Danish EPA and Miljoekontrollen, The Municipality of Copenhagen. The project has been carried out under the "Technology Demonstration and Evaluation Program".

REFERENCES

Bjerg, P.L., K. Rügge, J.K. Pedersen and T.H. Christensen. 1995. "Distribution of Redox Sensitive Groundwater Quality Parameters Downgradient of a Landfill (Grindsted, Denmark)." *Environmental Science & Technology*. 29(5): 1387-1394.

Brandt, V.P. 1998. "Reduction of various iron oxides with H_2. Investigation of the thermodynamics of the process." (in Danish). M.Sc.Thesis. Department of Geology and Geotechnical Engineering and Department of Environmental Science and Technology, Technical University of Denmark.

Chapelle, F.H., P.B. McMahon, N.M. Dubrovsky, R.F. Fujii, E.T. Oaksford and D.A. Vroblesky. 1995. "Deducing the distribution of terminal electron-accepting processes in hydrologically diverse groundwater systems." *Water Resources Research*. 31(2): 359-371.

Christensen, A.G. and C. Riis. 2000. "Technology demonstration project. Natural Attenuation of mineral spirits and chlorinated solvents in the ground water at Drejøgade 3-5." (in Danish). Miljøstyrelsen & Miljøkontrollen, Københavns Kommune. In press. (www.mst.dk).

Hoehler, T.M., M.J. Alperin, D.B. Albert and C.S. Martens. 1998. "Thermodynamic Control on Hydrogen Concentrations in Anoxic Sediments." *Geochimica et Cosmochimica Acta*. 62: 1745-1756

Hoehler, T.M. 1998. "Thermodynamics and the role of Hydrogen in anoxic sediments". Ph.D.Thesis. University of North Carolina.

Jakobsen, R., H.-J. Albrechtsen, M. Rasmussen, H. Bay, P.L. Bjerg and T.H. Christensen. 1998. "H_2 concentrations in a Landfill Leachate Plume (Grindsted, Denmark): In Situ Energetics of Terminal Electron Accepting Processes." *Environmental Science & Technology*. 32: 2142-2148.

Jakobsen, R. and D. Postma. 1999. "Redox zoning, rates of sulfate reduction and interactions with Fe-reduction and methanogenesis in a shallow sandy aquifer, Rømø, Denmark." *Geochimica et Cosmochimica Acta,* 63(1): 137-151.

Ludvigsen, L., H.-J. Albrechtsen, G. Heron, P.L. Bjerg and T.H. Christensen, 1998. "Anaerobic microbial redox processes in a landfill leachate contaminated aquifer (Grindsted, Denmark)." *Journal of Contaminant Hydrology.* 33: 273-291.

Parkhurst, D.L., D.C. Thorstenson and N.L. Plummer. 1980. "PHREEQE – A computer program for geochemical calculations". *U.S.G.S. Water Resour. Invest.* 80-96, 210.

Postma, D. and R. Jakobsen. 1996. "Redox zonation: Equilibrium constraints on the Fe(III)/SO_4-reduction interface". *Geochimica et Cosmochimica Acta.* 60(17): 3169-3175.

Schulz, S. and R. Conrad. 1996. "Influence of temperature on pathways to methane production in the permanently cold profundal sediment of Lake Constance." *FEMS Microbial Ecology* 20: 1-14.

Wiedemeier, T.H., M.A. Swanson, D.E. Moutoux, E.K. Gordon, J.T. Wilson, B.H. Wilson, D.H. Kampbell, J.E. Hansen, P. Haas, and F.H. Chapelle. 1996. "Technical Protocol for Evaluating Natural Attenuation of Chlorinated Aliphatic Hydrocarbons in Ground Water". For: Air Force Center for Environmental Excellence, Technology Transfer Division, Brooks Air Force Base, San Antonio, Texas.

CARBON ISOTOPIC FRACTIONATION OF CHLORINATED ETHENES DURING BIODEGRADATION: FIELD APPLICATIONS

G.F. Slater[1], B. Sherwood Lollar[1], E. Edwards[1], B. Sleep[1]
M. Witt[2], G.M. Klecka[2], M.R. Harkness[3], J.L. Spivack[3]
1. University of Toronto, Toronto, Canada. 2. The Dow Chemical Company, Midland, Michigan. 3. General Electric Research and Development Center, Schenectady, NY.

ABSTRACT: Compound Specific Isotope Analysis (CSIA) of organic compounds in the subsurface provides an efficient means of constraining subsurface processes by measuring changes in the stable carbon isotopic composition ($\delta^{13}C$) of individual compounds directly. Laboratory studies have demonstrated that for the chlorinated ethenes non-degradative processes, dissolution, sorption, volatilization, are isotopically conservative within 0.5‰. In contrast, reductive dechlorination of TCE by two enriched microbial consortia was strongly isotopically fractionating. The use of isotopic analysis to indicate the occurrence and effectiveness of natural attenuation was evaluated at Area 6, Dover Air Force Base (Dover, Delaware). Isotopic enrichment of TCE was observed in wells in the anaerobic portion of the contaminant plume, consistent with laboratory microcosm results. Using a mean enrichment factor for dechlorination of TCE from the laboratory results, it was estimated that isotopic enrichment of TCE in the most isotopically enriched well represented a biologically mediated depletion more than 50% relative to TCE in the proposed source area.

INTRODUCTION

The development of high sensitivity GC/C/IRMS (gas chromatograph - combustion - isotope ratio mass spectrometry), has allowed compound specific isotopic analysis (CSIA) of primary groundwater contaminants directly. CSIA is therefore a new tool for investigation of the source and fate of hydrocarbon contaminants. Initial studies focused on characterizing the isotopic signatures of contaminants from different sources in order to establish the potential range of source variation in $\delta^{13}C$ values (van Warmerdam et al. 1995, Dempster et al. 1997). Differentiation of contaminant sources using CSIA is possible if the $\delta^{13}C$ values of the contaminant are not altered while undergoing subsurface processes. Laboratory experiments were therefore carried out to constrain the kinetic isotope effects associated with key subsurface processes controlling contaminant fate and transport in groundwater.

Studies characterizing stable carbon isotope effects during non-degradative subsurface processes, i.e. sorption, volatilization, and dissolution, established that fractionation associated with these non-degradative processes are small, and in many cases less than analytical uncertainty (Dempster et al. 1997, Harrington et al. 1999, Slater et al. 1999, Poulson et al. 1999). In contrast, for chlorinated hydrocarbons, both abiotic (Slater et al. 1998, Dayan et al. 1999) and biodegradation reactions (Heraty et al. 1999, Hunkeler et al. 1999, Sherwood Lollar et al. 1999) have been shown to produce large and reproducible kinetic effects that suggest the potential to use stable carbon isotope analysis as a means of identifying and monitoring degradative processes. Due to preferential degradation of bonds containing the light (^{12}C) versus heavy isotope (^{13}C), isotopic enrichment of the residual contaminant in ^{13}C results in shifts in $\delta^{13}C$ values (fractionation) of more than 10 ‰ during reductive dechlorination of PCE and TCE (Hunkeler et al. 1999, Sherwood Lollar et al. 1999).

This study characterized isotopic fractionation factors for reductive dechlorination of TCE by two different enriched microbial consortia in laboratory batch studies. Then isotopic analysis was applied to PCE and TCE contamination at Area 6, Dover Air Force Base (Dover Delaware) to provide definitive evidence for the occurrence of reductive dechlorination. This report will focus only on the TCE results obtained from the site.

Site Description. Area 6 at Dover Air Force Base (Dover, Delaware) is part of the industrial portion of the base where solvents were commonly used for aircraft and vehicle maintenance, painting, stripping and welding. The majority of the contamination at the seven identified source areas consisted of chlorinated solvents, both PCE and TCE (Figure 1) (Klier et al. 1999). Since 1995, Area 6 has been the location of an in-depth study by the RTDF (Remediation Technologies Development Forum Bioremediation of Chlorinated Solvents Action Team) to evaluate the effectiveness of intrinsic bioremediation (natural attenuation) of the PCE and TCE contamination in groundwater and sediments.

Results of the geological, hydrogeological and biogeochemical study (Kleir et al. 1999), and computer modeling of the site (Clement et al. 2000) have been previously published. Since PCE and TCE were the only chlorinated solvents used in Area 6, the presence of daughter products (cis-DCE, VC and ethene) in the deep aquifer is indicative of biological reductive dechlorination. Additional lines of biogeochemical and microbiological evidence for intrinsic bioremediation via reductive dechlorination have been outlined by (Kleir et al. 1999).

METHODOLOGY:

Laboratory studies. Batch microcosm experiments to characterize stable carbon isotopic fractionation by two different microbial consortia were carried out at the University of Toronto. The two consortia used were the Pinellas consortium, generated from a TCE contaminated field site, and the KB-1 consortium, generated from a different TCE contaminated site. The details of the study of the Pinellas consortium have been previously presented in Sherwood Lollar et al (1999). For the KB-1 consortium, the experimental set up consisted of anaerobic 250 mL serum vials containing 100 mL of defined mineral media, 100 mg/L methanol as electron donor and 20 mg/L of TCE. Analysis of chlorinated ethene concentration involved injection of 300 μL of the vial headspace on a HP-5890 Gas Chromatograph (GC) equipped with a GS-Q column (J&W Scientific) and an FID detector. Isotopic analysis involved injection of 300 to 1000 μL of vial headspace into a GC/C/IRMS system consisting of a Varian 3400 GC equipped with a DB-

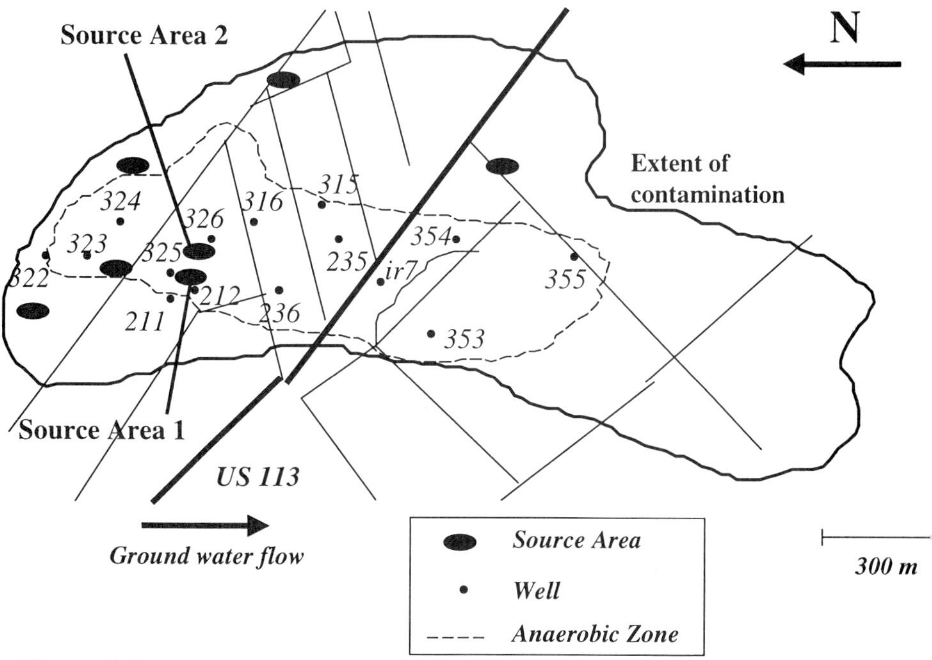

Figure 1. Proposed contaminant source zones at Area 6, Dover Air Force Base. The extent of the contaminant plume is shown in solid outline. The extent of the anaerobic zone of the plume is shown in dotted outline. Location of sampled wells and source zone 1 and 2 is also shown. Groundwater flow direction is to the south.

624 column connected directly to a Finnigan Mat 252 isotope ratio mass spectrometer. While internal reproducibility based on triplicate sample injection is generally between < 0.3 ‰, differences between samples (error bars) are assigned a value of no less than 0.5 ‰ to incorporate both reproducibility and accuracy of the measurement.

Field Study. Field samples were collected from 15 deep zone wells in the vicinity of two proposed source zones in Area 6, source area 1 and source area 2, and approximately upgradient and downgradient from these sources areas (Figure 1). Transect I refers to a sampling transect approximately downgradient and upgradient of source area 1 (wells 211/212), encompassing downgradient wells 236 and 353; and upgradient wells 325, 323, and 322. Transect II refers to a sampling transect approximately downgradient and upgradient from source area 2 (well 326), encompassing downgradient wells 316, 315, 235, ir7, 354, 355; and upgradient well 324. Samples were collected from each monitoring well using dedicated bladder pumps and Teflon™-lined polyethylene tubing. Field parameters (dissolved oxygen, pH, temperature, specific conductance and redox potential) were recorded until they stabilized, before separate samples were collected for isotopic and geochemical analysis.

Groundwater samples for geochemical analysis of organic constituents were analysed by Lancaster Laboratories, Lancaster, Pennsylvania. Volatile organic carbon (VOC) analyses were conducted in accordance with USEPA SW-846 method 8260A using gas chromatography and mass spectrometry.

Samples for isotopic analysis were collected as per standard VOC samples in glass VOA sample vessels (headspace free) fitted with Teflon-coated septa. Samples were kept cold (4°C), shipped to University of Toronto and analyzed within 1 week. Compound specific carbon isotope analysis of all chlorinated ethenes was done at the University of Toronto using the GC/C/IRMS system described for the laboratory studies.

Stable carbon isotope analysis involves measurement of the two stable isotopes of carbon, ^{12}C and ^{13}C to establish $\delta^{13}C$ signatures where:

$$\delta^{13}C = (Rsample/Rstandard - 1) \times 1000 \qquad \text{(in units of ‰)(1).}$$

Rsample is the $^{13}C/^{12}C$ ratio in a given compound, which is normalized to Rstandard, the $^{13}C/^{12}C$ ratio in a standard reference material, in this case V-PDB. The fractionation factor (α) between any two compounds a and b, is defined as:

$$\alpha = (1000 + \delta^{13}C_a)/(1000 + \delta^{13}C_b) \qquad (2).$$

Fractionation factors can be converted to permil enrichment factors (ε) such that $\varepsilon = 1000(\alpha-1)$. The enrichment factor is related to the fraction of substrate remaining (f) such that

$$\varepsilon = (\delta^{13}C_{initial} - \delta^{13}C)/\ln f \qquad (3)$$

RESULTS AND DISCUSSION

Laboratory studies. The results of the KB-1 consortium experiment were consistent with those observed for the Pinellas consortium in Sherwood Lollar et al (1999). The rate constant for TCE was 0.02 to 0.2 (hr^{-1}). Production and subsequent dechlorination of daughter products, cDCE and VC, was observed until production of ethene as a final product. The $\delta^{13}C$ of the TCE became more positive as dechlorination proceeded, consistent with preferential degradation of ^{12}C containing molecules over ^{13}C containing molecules. The relationship between isotopic composition and extent of degradation was modeled with a Rayleigh evolution model using an enrichment factor of -13.8 ‰ (Figure 2). The KB-1 data shown in Figure 2 represents data from four repetitions of this experiment. Variation in the rate of degradation between these repetitions had no observable impact on the relationship between isotopic fractionation and extent of degradation. The results of the previous experiment using the Pinellas consortium are also shown in Figure 2. It can been seen that the same excellent agreement

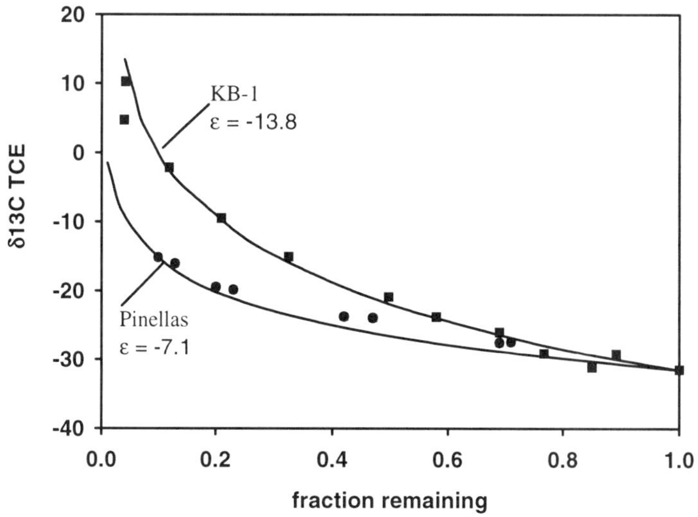

Figure 2. Isotopic composition versus fraction of TCE remaining for batch laboratory experiments degrading TCE with the KB-1 and Pinellas consortia. Enrichment factors determined for each culture are shown. Error bars representing 0.5 ‰ accuracy and reproducibility of GC/C/IRMS analysis are smaller than plotted symbols.

was observed between the data and the Rayleigh model, which in this case used an enrichment factor of -7.1 ‰. The consistent observation of isotopic enrichment during reductive dechlorination of TCE indicates that isotopic analysis can be used to qualitatively determine the occurrence of anaerobic biodegradation. Clearly, however, the results indicate there is a significant difference in the enrichment factor for the two different cultures. The critical question is whether this variability has a significant impact on the potential to use the degree of isotopic fractionation to quantitatively estimate the extent of biodegradation in the field.

Field study. Due to the absence of detectable free product on the site now, the original $\delta^{13}C$ signature of the solvent spilled at the various source areas was assumed to be that observed at the wells with the highest concentrations of dissolved TCE nearest the source zones which should be least affected by processes of biodegradation. The well nearest source area 2 (326) and the two wells in source area 1 (211/212) do indeed have some of the least enriched (most negative) $\delta^{13}C$ signatures at the site at -26.0 ‰ and -25.0 ‰ respectively.

Figure 3 shows concentrations and the $\delta^{13}C$ values for TCE for all the wells sampled in the course of the study. All wells other than those mentioned above are located away from the source zones and if as hypothesized, biodegradation is primarily responsible for decreasing concentration levels of chlorinated ethenes at the site, then the $\delta^{13}C$ values of these wells should show evidence of isotopic enrichment related to that biodegradation. The values obtained for $\delta^{13}C_{TCE}$ at the site show definitive evidence for biodegradation effects as hypothesized. For both Transect I and Transect II, with the exception of well 235, all downgradient wells (n=8) are enriched in ^{13}C versus their respective source zone wells. In addition, $\delta^{13}C_{TCE}$ values for the upgradient wells (323, 324, 322) are enriched by several ‰ with respect to the source area wells. It is particularly notable that the well (236) with the most dramatically enriched $\delta^{13}C$ value (-18.0 ‰) is indeed the most biologically active well at the site based on it having the highest concentrations of all the biological daughter products of reductive dechlorination.

Notably, the upgradient wells (324, 323, 322) have among the most isotopically enriched $\delta^{13}C$ values at the site (Figure 3). Dispersive modeling results suggest that PCE, TCE and cis-DCE concentrations in this upgradient zone cannot be solely attributed to dispersive transport from source zones 1 and 2

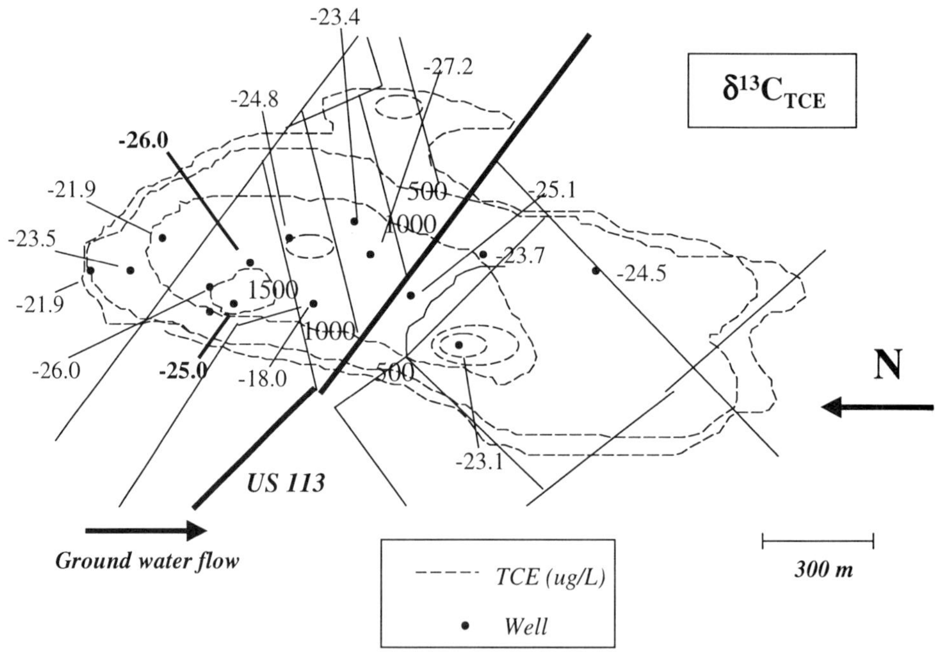

Figure 3. Concentration contours for TCE at Area 6 are shown in dotted lines. Isotopic compositions of TCE from each sampling well are shown. Isotopic composition of wells nearest source zones are shown in bold.

against the direction of advective flow. These higher than predicted concentrations of chlorinated ethenes indicate that there may be an additional source of contaminant migrating into the area of wells 323, 324, and 322. While the isotopic data do not necessarily confirm such a scenario, they are consistent with this hypothesis. What the isotopic results definitely indicate is that whatever the source, TCE in this area of the site is significantly biodegraded with respect to the source zone 1 and 2 wells and likely the most biodegraded in the study area.

Quantification of Extent of Biodegradation. As stated, the enrichment factor of -13.8 ‰ calculated for the dechlorination of TCE by the KB-1 consortium is significantly different that the value of -7.1‰ reported by Sherwood Lollar et al. (1999) for the Pinellas consortium. This suggests that the magnitude of the fractionation factor for microbial reductive dechlorination of chlorinated ethenes will be a function of the specific population of microbial dechlorinators. For the Dover AFB site then, the most accurate quantitative estimate of the amount of biodegradation (f) using changes in the $\delta^{13}C$ values of TCE between source wells and any other well requires that the enrichment factor (ε) for the specific microbial population at the Dover AFB site be determined by experimentation. In the absence of such information however, at least an estimate of the extent of biodegradation can be determined using the ε values determined from the laboratory experiments. Using Equation 4, and the values of source well 212 (-25.0 ‰) and of downgradient well 236 (-18.0 ‰) for $\delta^{13}C_{initial}$ and $^{13}C_{TCE}$, we obtain estimates of approximately 40% and approximately 64% respectively for the two values of ε. This estimate assumes that the TCE at well 236 is indeed derived from that in source area 1 (212) and therefore that the isotopic variation between the two wells is due to biodegradation and not mixing with a different source of TCE. Accurate quantification of the extent of biodegradation between wells at a given site will clearly require i. characterization of the appropriate fractionation factor for the in situ microbial population of dechlorinators and ii. comparison of the isotopic method with groundwater flow and transport model estimates of the extent of biodegradation. Nonetheless, the approach is already able to provide an estimate of the relative extent of biodegradation between zones of the contaminant plume. Despite the uncertainty in the enrichment factor, both estimates (-13.8‰ and -7.1‰), indicate that TCE at well 236 is > 40% more degraded than that in the proposed source area 1 (well 211/212).

At many field sites such as Dover AFB, where significant TCE still exists, evaluating the extent of biodegradation will be dependent on interpretation of the fractionation effects associated with the primary contaminants, PCE and TCE. As this study shows, this can be successfully done. Even at a complex site such as Area 6, Dover AFB, with multiple source areas and with the complicating factor of mixing between different contaminant sources, the observed variation in $\delta^{13}C$ values of PCE and TCE can be used to help constrain hypotheses concerning source zones, and to definitively identify zones of biodegradation. While estimates of the relative extent of biodegradation between different wells can be obtained, the ultimate goal for this approach should be obtaining a higher level of accuracy in quantification by increasing the existing database on specific fractionation factors associated with different microbial dechlorinating populations through continued laboratory studies, and by integrating the stable isotope constraints into groundwater transport models.

REFERENCES

Clement, T. P., C. D. Johnson, Y. Sun, G. M. Klecka and C. Bartlett. 2000. "Attenuation of Chlorinated Ethene Compounds: Model Development and Field-Scale Application at the Dover Site." *Journal of Contaminant Hydrogeology*. 42(113-140).

Dayan, H., T. Abrajano, N. C. Sturchio and L. Winsor. 1999. "Carbon Isotopic Fractionation During Reductive Dehalogenation of Chlorinated Ethenes by Metallic Iron." *Organic Geochemistry*. 30(755-763).

Dempster, H. S., B. Sherwood Lollar and S. Feenstra. 1997. "Tracing Organic Contaminants in Groundwater: A New Methodology Using Compound-Specific Isotopic Analysis." *Environmental Science and Technology*. 31(11): 3193-3197.

Harrington, R. R., S. R. Poulson, J. I. Drever, P. J. S. Colberg and E. F. Kelly. 1999. "Carbon Isotope Systematics of Monoaromatic Hydrocarbons: Vaporization and Adsorption Experiments." *Organic Geochemistry*. 30(765-775).

Heraty, L. J., M. E. Fuller, L. Huang, T. Abrajano Jr. and N. C. Sturchio. 1999. "Isotopic Fractionation of Carbon and Chlorine by Microbial Degradation of Dichloromethane." *Organic Geochemistry*. 30(793-799).

Hunkeler, D., R. Aravena and B. J. Butler. 1999. "Monitoring Microbial Dechlorination of Tetrachloroethene (PCE) in Groundwater Using Compound-Specific Stable Carbon Isotope Ratios: Microcosm and Field Studies." *Environmental Science and Technology*. 33(16): 2733-2738.

Klier, N. J., G. M. Klecka, E. J. Lutz, D. E. Ellis, F. H. Chapelle and M. E. Witt. 1999. The Groundwater Geochemistry of Area 6, Dover Air Force Base, Dover, Delaware

Poulson, S. R., and J. I. Drever. 1999. "Stable Isotope (C, Cl, and H) Fractionation during Vaporization of Trichloroethylene." *Environmental Science and Technology*. 33(20): 3689-3694.

Sherwood Lollar, B., G. Slater, J. Ahad, B. Sleep, J. Spivack, M. Brennan and P. MacKenzie. 1999. "Contrasting Carbon Isotope Fractionation During Biodegradation of Trichloroethylene and Toluene: Implications for Intrinsic Bioremediation." *Organic Geochemistry*. 30(813-820).

Slater, G. F., H. D. Dempster, B. Sherwood Lollar, J. Spivack, M. Brennan and P. MacKenzie. 1998. "Isotopic tracers of degradation of dissolved chlorinated solvents." First International Battelle Conference on Remediation of Chlorinated and Recalcitrant Compounds. Monterey, CA.

Slater, G. F., H. S. Dempster, B. Sherwood Lollar and J. Ahad. 1999. "Headspace Analysis: A New Application for Isotopic Characterisation of Dissolved Organic Compounds." *Environmental Science and Technology*. 33(1): 190-194.

van Warmerdam, E. M., S. K. Frape, R. Aravena, R. J. Drimmie, R. J. Flatt and J. A. Cherry. 1995. "Stable Chlorine and Carbon Isotope Measurements of Selected Chlorinated Organic Solvents." *Applied Geochemistry*. 10(547-552).

MODELING RECALCITRANT INTERMEDIATE FORMATION DURING BIODEGRADATION

J.M. VanBriesen (Carnegie Mellon University, Pittsburgh, PA)
B.E. Rittmann (Northwestern University, Evanston, IL)

ABSTRACT: Complex biological degradation reactions proceed through intermediates. Recalcitrant intermediates sequester carbon and electrons, reducing cell growth and slowing the expected rate of contaminant removal. The distribution of reducing equivalents and energy between catabolic and anabolic processes in the microorganism control the bacterial yield. Bacterial yields and step-wise stoichiometry for a biodegradation pathway can predict the relationships among primary electron-donor and acceptor substrates, degradable and recalcitrant intermediates, and degradation products (e.g., H_2CO_3, NH_4^+, or H^+). The stoichiometry of the multi-step degradation pathway is coupled to a kinetic biological reaction model that predicts the rate of substrate removal, cell growth, and the changes to all chemicals affected by the loss of primary substrate and formation and persistence of intermediates. A methodology for bacterial yield predictions and stoichiometric formulations for compounds degraded through intermediates is presented. Degradation of the anthropogenic chelating agent, nitrilotriacetate (NTA), is presented as an example.

INTRODUCTION

Biological modeling is designed to predict the behavior of reactions catalyzed by microorganisms. Microbial growth requires (1) nutrients (e.g., carbon and nitrogen) to build cell structures, (2) electrons to reduce nutrients to the oxidation state of cell materials, and (3) energy to drive synthesis reactions. Microorganisms often acquire nutrients, electrons, and energy from a single substance. Oxidation of this primary substrate provides electrons and carbon to the cells. When the primary electron-donor substrate is fully mineralized, the maximum microbial growth occurs. But, if the substrate is converted to recalcitrant intermediates, these compounds sequester reducing equivalents, lowering the overall yield of cells. The lower concentration of cells slows the utilization of the primary substrate and can increase the time required to remove the original contamination. In addition to the delayed removal of the primary substrate, intermediates may persist and pose a long-term problem.

BIOLOGICAL MODELING METHODOLOGY

Cell yield (Y) is a fundamental biological parameter that allows prediction of the full stoichiometry of biological growth/substrate utilization reactions and thus is necessary for modeling coupled biogeochemical systems. The yield is combined with mass balances (McCarty 1971; McCarty 1972; McCarty 1975; Heijnen and vanDijken 1992; Heijnen et al 1992; Heijnen 1994; VanBriesen and Rittmann 2000a) to predict the overall stoichiometry.

Often the overall yield for the oxidation of an organic electron-donor substrate can be measured experimentally. However, for compounds that undergo degradation through intermediates that may persist, experimental yield determination may be infeasible (Heijnen and vanDijken 1992; VanBriesen and Rittmann, 2000a). A systematic method for *predicting* yields in biological degradation reactions is required.

Yields for microbial reactions that proceed through intermediate-formation are predicted based on the assumption that electron and energy balances control microbial growth. Figure 1 shows this schematically. The primary electron donor is oxidized to the intermediate, releasing electrons (T) that are sent to the electron acceptor for energy generation (f_e^0) and to the nutrient sources for reduction to the level necessary to grow biomass (f_s^0). For each step in the degradation pathway, some electrons are held in intermediates (H), thus reducing the electrons available for energy generation or cell synthesis.

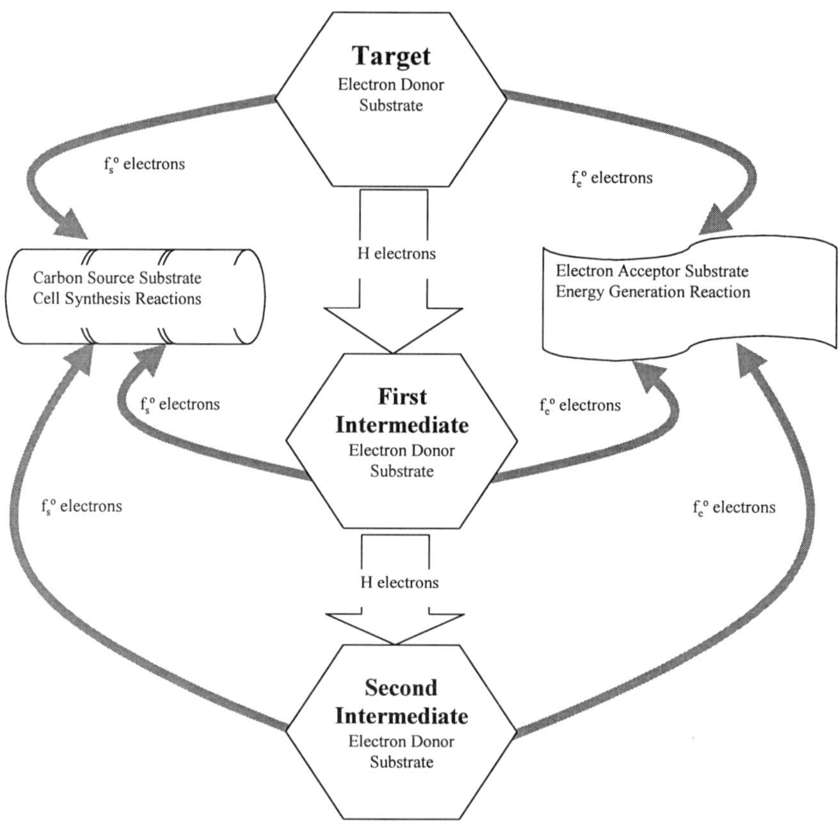

FIGURE 1. Distribution of Electrons to Energy Generation and Cell Synthesis

Mathematically, the energy and electron balances are represented as:

$$-f_e^o K\Delta G_R^{o'} = f_s^o \Delta G_{syn}^{o'} \tag{1}$$

$$f_e^o + f_s^o = T \;\; ; \;\; T + H = 1 \tag{2}$$

in which $\Delta G_R^{o'}$ is the standard free energy of the energy generating redox couple, $\Delta G_{syn}^{o'}$ is the standard free energy of the cell synthesis reaction, K is the efficiency of energy capture in the energy-generation reaction, f_e^o is the fraction of electron-donor electron equivalents sent to the acceptor to drive the energy generating redox reaction, f_s^o is the fraction of electron-donor electron equivalents invested in biomass via the synthesis reaction, T is the fraction of electron-donor electron equivalents that are released from the electron-donor in the step, and H is the fraction of electron-donor electron equivalents that are held in the intermediates. Equation 1 says that the energy made available from transfer of electrons from the donor to the acceptor ($-f_e^o K\Delta G_R^{o'}$) is invested to synthesize f_s^o electron equivalents of biomass ($f_s^o \Delta G_{syn}^{o'}$). Equation 2 says that the number of electrons sent to synthesis and energy generation cannot exceed the number of electrons available from the partial oxidation of the electron donor in the step and that the transferred and held electron fractions total 1.

Knowledge of the structure of the formed intermediate is required to determine T and H. For example, Figure 2 shows the degradation of nitrilotriacetate (NTA) through its intermediates: iminodiacetate (IDA), glycine, and glyoxylate. In the initial step (conversion of NTA to IDA and glyoxylate), 2 electrons are released (T=2) while 16 are held (H=16) in the intermediates (12 in IDA and 4 in glyoxylate).

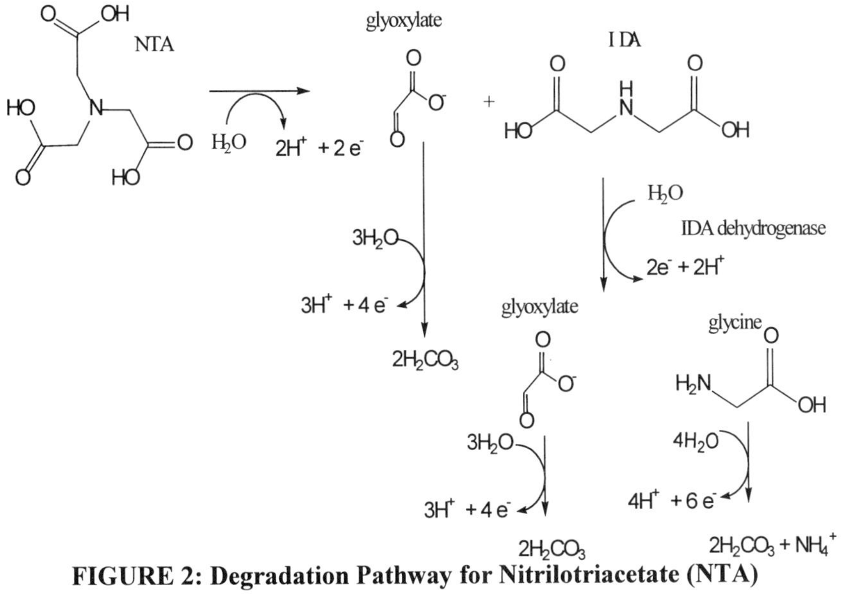

FIGURE 2: Degradation Pathway for Nitrilotriacetate (NTA)

Once T is known, coupling equations (1) and (2) numerically allows solution for the unknown f_s^o and f_e^o. The yield is then predicted from:

$$\left(\frac{Y\,moles\ cells}{mole\ substrate}\right) = f_s^o \left(\frac{mole\ cells}{EQC\ e^-\ equivalents}\right)\left(\frac{EQS\ e^-\ equivalents}{mole\ substrate}\right) \quad (3)$$

where EQS is the electron equivalents available per mole of substrate and EQC is the electron equivalents required to synthesize cells.

The f_s^o, f_e^o, and T values are combined with half reactions for cell synthesis, electron donor oxidation and electron acceptor reduction, following:

$$Overall\ \text{Re}action = -TR_d + f_e^o R_a + f_s^o R_c \quad (4)$$

where R_d is the half reaction for the electron donor transformation to intermediate, R_a is the half reaction for the electron acceptor, and R_c is the half reaction for cell synthesis. Equation 4 states that electrons released from the electron donor in the degradation step (T) are sent to the acceptor for energy generation and to the cell synthesis reaction for cell formation in proportions determined by the yield.

The cell synthesis half reaction (R_c) can be created based on the cell formula (VanBriesen and Rittmann, 2000a). For example, using a cell formula of $CH_2O_{0.6}N_{0.2}$ (degree of reductance, γ_c=4.2; carbon fraction, σ_c=0.454), the synthesis half reaction based on ammonia as a nitrogen source is:

$$0.238\ H_2CO_3 + 0.0476\ NH_4^+ + 0.952\ H^+ + e^- \rightarrow 0.238\ CH_2O_{0.6}N_{0.2} + 0.571\ H_2O \quad (5)$$

The electron acceptor half reaction (R_a) for molecular oxygen is:

$$1/4\ O_2 + e^- + H^+ \rightarrow 1/2\ H_2O \quad (6)$$

Generating the electron donor half reaction (R_d) is more complex and depends upon the formation of intermediates, the use of cosubstrates, and the oxidation state of any released nitrogen. McCarty(1975) presents a systematic methodology for electron donor substrates that are mineralized in the absence of activation reactions. VanBriesen and Rittmann (2000a) present a systematic methodology for electron donor substrates that are oxidized to intermediates in the absence of activation reactions. For the NTA degradation system described by Figure 2, the stepwise degradation reactions, written in standard reduction format for one electron equivalent are:

$$½\ C_4H_5O_4N^{2-} + ½\ C_2HO_3^- + H^+ + e^- \rightarrow ½\ C_6H_6O_6N^{3-} + ½\ H_2O \quad (7)$$

$$½\ C_2HO_3^- + ½\ C_2H_4O_2N^- + H^+ + e^- \rightarrow ½\ C_4H_5O_4N^{2-} + ½\ H_2O \quad (8)$$

$$1/3\ H_2CO_3 + 2/3\ H^+ + 1/6\ NH_4^+ + e^- \rightarrow 1/6\ C_2H_4O_2N^- + 2/3\ H_2O \quad (9)$$

$$½\ H_2CO_3 + ¾\ H^+ + e^- \rightarrow ¼\ C_2HO_3^- + ¾\ H_2O \quad (10)$$

The prediction of the overall stoichiometry (following equation 4) for each step in the degradation pathway allows for coupling microbiological reactions and chemical reactions within a biogeochemical model. Rittmann and VanBriesen (1996) and VanBriesen and Rittmann (2000b) describe CCBATCH, a model designed to describe interactions between aqueous geochemical reactions and biological degradation reactions. The rate of biological reactions is described by the dual Monod kinetic formulation:

$$r_{util} = -q_m X_a \frac{S}{S+K_S} \frac{A}{A+K_A} \tag{11}$$

in which r_{util} is the rate of utilization of the rate-limiting substrate ($M_sL^{-3}T^{-1}$), q_m is the maximum specific rate of substrate utilization for the bacterial species ($M_sM_x^{-1}T^{-1}$), X_a is the concentration of metabolically active organisms in the system (M_XL^{-3}), S is the concentration of the electron-donor substrate (M_sL^{-3}), K_s is the half-maximum-rate concentration for the electron-donor substrate (M_sL^{-3}), A is the concentration of the electron-acceptor substrate (M_aL^{-3}), and K_A is the half-maximum concentration of the electron-acceptor substrate (M_aL^{-3}).

In CCBATCH, the rate of substrate utilization (determined following equation 11) is coupled to the stoichiometry reaction (determined following equation 4), which predicts the relative change in all species affected by the biodegradation. This leads to a suite of rates, one for each component affected by the biological reaction. These rates are used to adjust the total concentration of each component and the aqueous system is then re-speciated.

MODELING EXAMPLE

NTA biodegradation has been widely observed in natural and laboratory systems, and it typically results in production of mineralization products: H_2CO_3 and NH_4^+ (Warren and Malec 1972; Tiedje et al. 1973) However, under some conditions, complete loss of NTA does not result in all of the carbon in NTA being released as H_2CO_3 or incorporated into cells. Instead, a pool of dissolved organic carbon consisting of biomass decay products and degradation intermediates persists in solution following the complete elimination of NTA (Bolton et al. 1996; Quinn 1996). The NTA system, as described by Figure 2 and equations (7-10) is presented as an example of the potential impact of intermediate formation on the removal of the primary substrate. Following VanBriesen and Rittmann (2000a), we combine equation 8 and 9 by assuming glycine mineralization is rapid and accounts for the rapid production of ammonia in solution during the degradation of NTA. Equation 11 is used to predict the rate of substrate utilization for NTA, IDA, and glyoxylate based on literature kinetic values. For NTA degradation, the maximum rate of substrate utilization (q_{max}) is reported as 0.178 mol NTA/mol cells-hour (Wanner et al. 1990), and the Monod half-maximum rate constant for the electron donor (K_s) is reported as 7.85 x 10^{-7} mol NTA/L (Siegrist et al. 1989). For the simulations presented here, the K_s value for IDA and glyoxylate are assumed to be equal to the value for NTA. We vary q_{max} for IDA and glyoxylate from 1/2 the value for NTA to twice the value for NTA to evaluate the impact of different kinetics of intermediate degradation on the loss of the primary substrate (NTA) in solution. All the simulations shown here are based on an assumed starting cell concentration of 10^{-6} M, a starting NTA concentration of 10^{-3} M, and a system pH of 7.0.

Figure 3 shows the results of four simulations for concentration of substrates. The predicted NTA profile based on assuming one-step mineralization and neglecting intermediate formation is shown on each panel for comparison. Panels a-c consider cases where one or both of the intermediates has a maximum

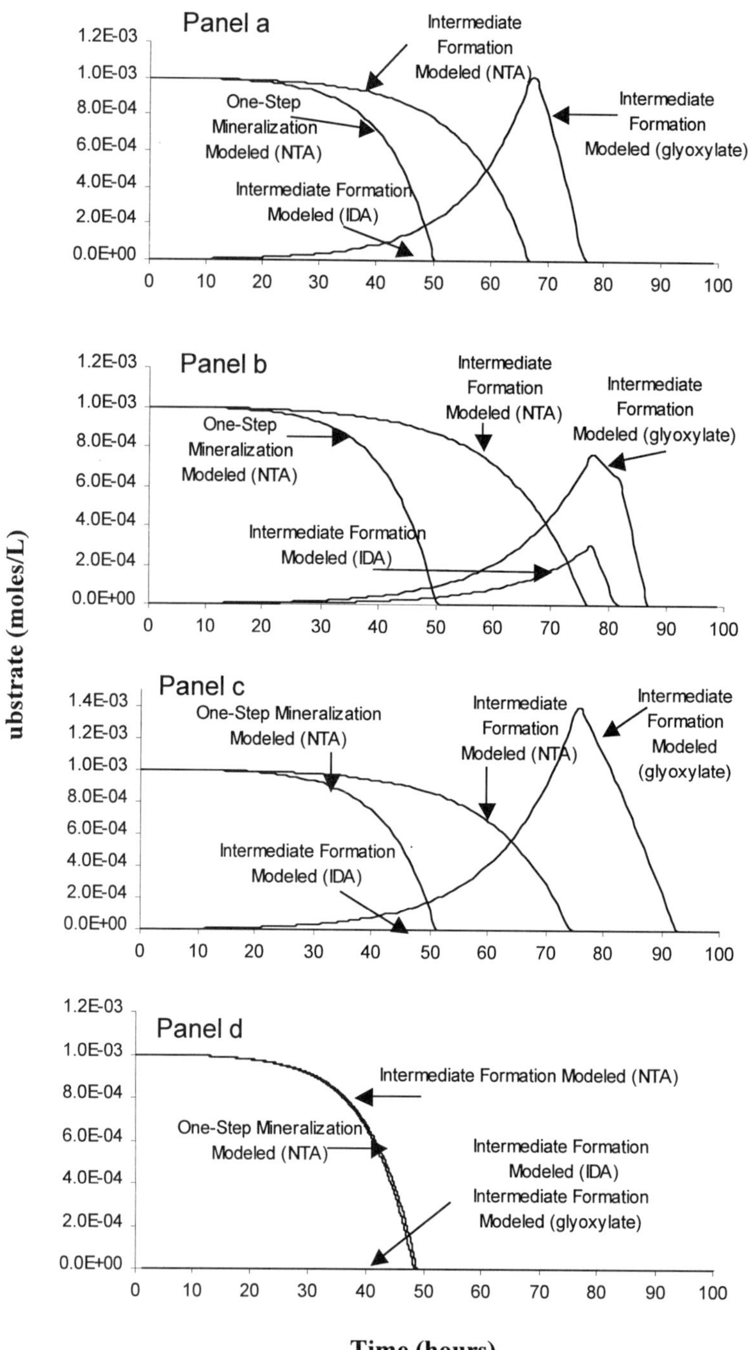

Figure 3. Intermediate Modeling Simulations with NTA Biodegradation

rate of substrate utilization equal to or less than the rate of utilization for the primary substrate. Table 1 gives the assumed q_{max} values for each simulation.

TABLE 1. Maximum Specific Rate of Substrate Utilization for Simulations

	Panel a	Panel b	Panel c	Panel d
q_{max} NTA	0.178	0.178	0.178	0.178
q_{max} IDA	0.178	0.089	0.178	0.356
q_{max} glyoxylate	0.178	0.178	0.089	0.356

The first case (panel a) involves assuming the reactions in the three-step pathway have the same maximum substrate utilization rate. In step one (reaction 7; NTA to IDA and glyoxylate) carbon and electrons are sequestered in the intermediates rather than released for cell growth. With the delay of cell synthesis, the substrate utilization rate is slowed. NTA is removed from solution by 68 hours compared with 50 hours when intermediate formation is neglected.

Panels b and c show cases where the IDA and glyoxylate rates are assumed to be slower than other utilization rates in the system. In panel b, the slower rate of IDA utilization means more carbon and electrons are held in this intermediate and not available for cell synthesis and energy generation reactions. The slower growth of cells leads to a greater delay in removal of NTA (78 hours compared with 50 when intermediates are ignored and 68 when all rates are taken as equivalent). In panel c, the slower rate of glyoxylate utilization delays the removal of NTA from solution but in this case IDA does not accumulate since its utilization rate is faster. The slower glyoxylate utilization rate causes this intermediate to accumulate in excess of the original NTA concentration (recall 2 moles of glyoxyate are produced for each mole of NTA degraded). The complete removal of all substrates is delayed till beyond 90 hours for this case.

Finally (in panel d), the rate of utilization of the intermediates is assumed to be twice the rate of utilization of the primary substrate (NTA). The intermediates do not appear in solution, and the loss of primary substrate is identical to the case where one-step mineralization is postulated. The faster rates of utilization for the intermediates means formation of IDA and glyoxylate in step 1 is followed by instantaneous removal, mimicking a single step reaction.

CONCLUSION

Biological degradation of anthropogenic compounds often proceeds through multiple steps, and modeling these reactions requires a systematic method to develop reaction stoichiometries and predict intermediate yields. Formation of intermediates sequesters carbon and electrons away from cell synthesis and energy generation pathways and delays overall removal of primary contaminants and intermediates. Coupling an equilibrium-kinetic model, CCBATCH, with stoichiometries for the reaction steps allows prediction of the impact of different substrate utilization rates on the removal of contaminants and the transient formation and removal of intermediates.

REFERENCES

Bolton, H. Jr., D. C. Girvin, A. E. Plymale, S. D. Harvey, and D. J. Workman. 1996. Degradation of Metal-Nitrilotriacetate (NTA) Complexes by Chelatobacter heintzii. *Environmental Science and Technology* 30, no. 3: 931-38.

Heijnen, J. J., and J. P. vanDijken. 1992. In search of a thermodynamic description of biomass yields for the chemotrophic growth of microorganisms. *Biotechnology and Bioengineering* 39: 833-58.

Heijnen, J. J., M. C. M. vanLoosdrecht, and L. Tijhuis. 1992. A black box mathematical model to calculate auto- and heterotrophic biomass yields based on Gibbs energy dissipation. *Biotechnology and Bioengineering* 40: 1139-54.

Heijnen, JJ. 1994. Thermodynamics of microbial growth and its implications for process design. *Trends in Biotechnology* 12: 483-92.

McCarty, P. L. 1971. Energetics and Bacterial growth. *Organic Compounds in Aquatic Environments.* Editors S. D. and Hunter J. V. Faust New York: Marcel Dekker, Inc.

———. 1972. Energetics of Organic Matter Degradation. *Water Pollution Microbiology.* Editor Ralph Mitchell New York: Wiley-Interscience.

———. 1975. Stoichiometry of Biological Reactions. *Progress in Water Technology* 7: 157-72.

Quinn, J. W. 1996. Monitoring reactants and products of the aerobic degradation of nitrilotriacetic acid by *Chelatobacter heintzii Strain ATCC 29600.*Northwestern University Master Research Project Report.

Siegrist, H., A. Alder, W. Gujer, and W. Giger. 1989. Behaviour and Modelling of NTA Degradation in Activated Sludge Systems. *Water Science Technology* 21: 315-24.

Tiedje, JM, BB Mason, CB Warren, EJ Malec. 1973. Metabolism of nitrilotriacetate by cells of pseudomonas species. Appl. Microbiol 25: 811-813.

VanBriesen, J. M., and B. E. Rittmann. 2000a. Mathematical Description of Microbiological Reactions Involving Intermediates. *Biotechnology and Bioengineering* 67: 1: 35-52.

VanBriesen, J.M. and B.E. Rittmann. 2000b. Modeling Speciation Effects on Biodegradation in Mixed Metal/Chelate Systems. Accepted in *Biodegradation.*

Wanner, Ursula, Judith Kemmler, Hans-Ulrich Weilenmann, Thomas Egli, Tarek El-Banna, and Georg Auling. 1990. Isolation and growth of a bacterium able to degrade nitrilotriacetic acid under denitrifying conditions. *Biodegradation* 1: 31-41.

Warren, Craig B., and Edward J. Malec. 1972. Biodegradation of Nitrilotriacetic acid and related imino and amino acids in river water. *Science* 176: 277-78.

SIMULATIONS OF CONTAMINATION ATTENUATION BY NATURAL DILUTION AND FLUSHING

Hubao Zhang (Duke Engineering & Service, Albuquerque, New Mexico)
David M. Peterson (Duke Engineering & Service, Albuquerque, New Mexico)
Robert G. Knowlton (Decision FX, Inc. Bosque Farms, New Mexico)
Don Metzler (U.S. Department of Energy, Grand Junction, Colorado)

ABSTRACT: This paper examines the reduction of contamination by natural dilution and flushing in an aquifer. In process models, it is assumed that the contaminant is leached from a source near the ground surface, transported through the underlying unsaturated zone and subsequently transported in the saturated zone. Statistical packages provide probabilistic parameter input to the process models and the post-conditioning of simulation results to honor site data. Sensitivity analyses are performed by perturbing input parameters in the flow and transport models. Dilution and flushing effects are studied through changes of recharge rate and stream-aquifer interactions. Simulations indicate that losing streams decrease the contaminant concentration somewhat depending upon the unit length loss rate. For a Type I gaining stream, the dilution effect is only limited by the ratio of stream flow rate and aquifer discharge rate. For a Type II gaining stream, the dilution effect also depends on the hydraulic head difference between the stream and the aquifer, and the hydraulic conductivity of the streambed. The contaminant attenuation methodology is applied to a Department of Energy (DOE) site.

INTRODUCTION

Natural attenuation may include biodegradation, dilution, sorption, volatilization, and other chemical processes. In recent years, natural attenuation research has been concentrated on the biodegradation of organic chemicals (Rafai et al, 1987; Chapelle et al, 1996). However, natural dilution has also been employed as an alternative strategy, particularly when the contaminants of concern have little or no biodegradation potential (e.g., metals). These processes have been recognized to assist in site remediation (Metzler et al., 1997) but factors influencing their effectiveness have not been studied in a systematic fashion. The objective of this study is to use model simulations to assess the relative contribution of area recharge and stream aquifer interaction to contaminant attenuation in ground water.

The models used include a one-dimensional flow and transport model in the unsaturated zone and a three-dimensional finite-difference transport model in the saturated zone. Both models are operated in a probabilistic manner using Monte Carlo simulations. A hydraulic conductivity field may be generated and automatically input in the finite-difference transport model by specifying a random field structure to account for the spatial variability in addition to standard input of hydraulic conductivity layer by layer. Post-conditioning is then implemented by calculating the residuals between measured ground water concentrations and calculated concentrations from Monte Carlo realizations. If the residual from a model run is smaller than a specified root-mean-squared error (RMSE) criterion, the realization passes the conditioning test and is subsequently used for the prediction of contaminant transport in future times; otherwise the simulation results are rejected.

SIMULATIONS

Contaminant dilution via recharge to a shallow aquifer is clearly mechanism that can lead to plume attenuation. The relationship between the aquifer and an adjacent stream can also be very important when assessing the potential for contaminant attenuation. Dilution in the aquifer occurs when stream water mixes with it. Alternatively, the aquifer may pass its contaminants to surface water media. In this paper, we assume three types of stream-aquifer interactions may be possible: losing, gaining, and segmented streams. A losing stream is a stream that loses water to the aquifer, while a gaining stream is a stream that gains water due to ground water discharging from an aquifer. A segmented stream is a stream consisting of a number of segments, each of which can lose or gain water. In general, both gaining and losing streams can help to reduce the contaminant concentrations in an aquifer.

Simulation of dilution by area recharge

A one-dimensional horizontal column is used to simulate contaminant dilution by area recharge. The column consists of 50 blocks, each with 10m by 10m by 10m sides. A constant head is specified at the downstream end of the column and a flux rate of 100 m^3/yr is applied at the upstream end. In addition, a uniform recharge is applied on the top of the column through its entire length. Conceptually, the concentration of the contaminant is reduced as the recharge water is added to the flow tube. A hydraulic conductivity of 100 m/yr and a porosity of 0.35 are assigned to the column. A longitudinal dispersivity of 20 m is assumed. Both continuous and pulse contaminant sources are considered for recharge rates of 0.2, 0.1, 0.01, 0.001 m/yr. For comparison purposes, contaminant transport without recharge is also simulated. For a continuous source of concentration 1 mg/L, mass transport over a period of 100 years is simulated. For a pulse source, the source concentration is assumed to be 1 mg/L for the first 35 years and zero afterward.

Figures 1(a) and 1(b) show the contaminant concentration profiles along the x direction for the continuous- and pulse-sources, respectively, at a simulation time of 100 years. Both diagrams show that the overall magnitude of concentration is reduced as the recharge rate is increased. However, the concentration farther away from the source and close to the column's downstream end may be higher for a higher recharge rate because the recharge increases the ground water velocity of the mass transport. For the pulse source and recharge rates of 0.1 and 0.2 m/yr, the bulk of contaminant mass is removed from the column after 100 years. Figures 1(c) and 1(d) show the breakthrough curves in the last block of the column for continuous and pulse sources, respectively. Both sets of results show earlier breakthrough times and lower overall concentrations for a higher recharge rate.

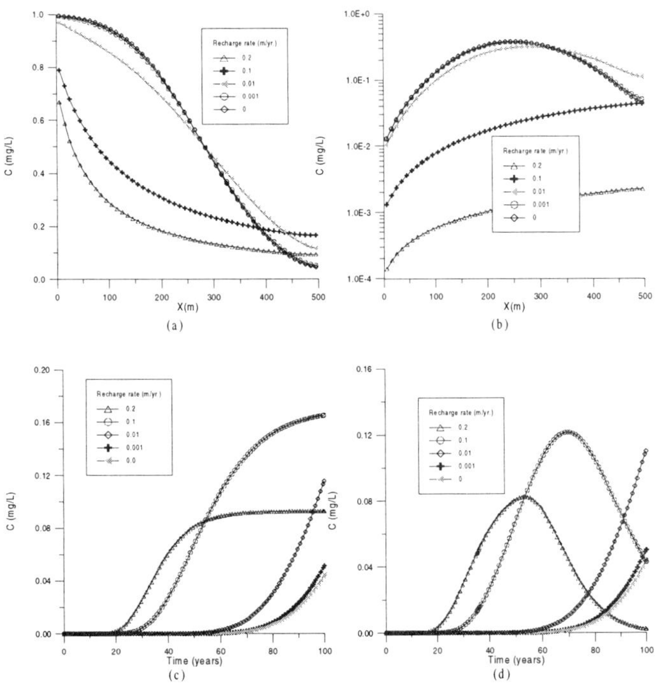

Figure 1. Simulations of dilution effects of recharge. (a) Concentration along the column for a continuous source; (b) Concentration along the column for a pulse source; (c) Breakthrough curve at x =500 m for a continuous source; (d) Breakthrough curve at x =500 m for a pulse source.

Simulation of contaminant attenuation by gaining streams

For descriptive purposes, a gaining stream that completely absorbs the contaminated water discharge from an aquifer is referred to as a Type I gaining stream, whereas a gaining stream that partially absorbs the contamination is called a Type II gaining stream. In this simulation, the simulated media include the source, unsaturated, and saturated zones. A source size of 140x160x8 m^3 is assumed near the ground surface in the unsaturated zone. An infiltration rate of 0.4 m/yr is applied to the source, which releases the contaminant through a sorption-limited process (Knowlton et al., 1998). The dissolved contaminant is subsequently transported across a 10-m thick unsaturated zone and downward to the saturated zone that is 15-m thick. Hydraulic conductivities of 1000, 6000, and 6000 m/yr are assigned to the source, unsaturated, and saturated zones, respectively. The van Genuchten model for a pressure-saturation-permeability relation is applied by assuming α=15 (1/m) and n=2.53 in the unsaturated zone. A horizontal hydraulic gradient of 0.003 is assigned uniformly to the regional flow field in the saturated zone. Porosities of 0.5, 0.43, and 0.34 are assigned to the source, unsaturated, and saturated zones, respectively. Dispersivity in the unsaturated model is given a value of 1 m because of the fine grid discretization in this zone. In the saturated zone, the longitudinal dispersivity is 20 m, and a longitudinal to transverse dispersivity ratio of 5 is assumed. Contaminant transport with 50 years of continuous waste release is simulated in this exercise.

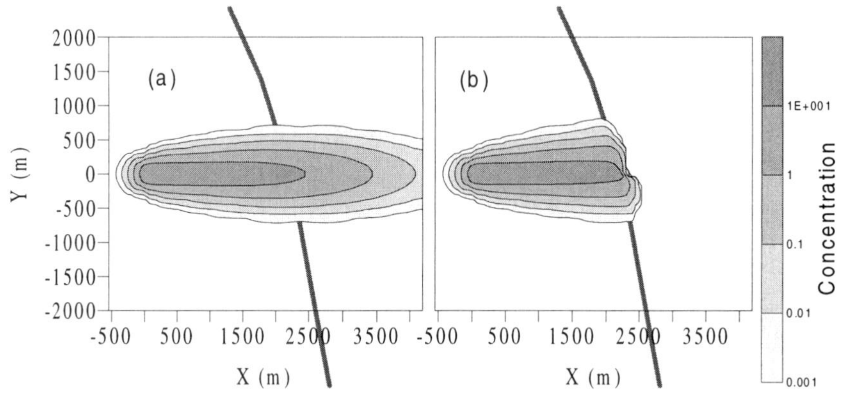

Figure 2. Simulation of dilution effects by a type I gaining stream. (a) concentrations without a stream; (b) concentrations with a type I gaining stream.

Figures 2(a) and 2(b) show examples of respective concentration distributions with and without a Type I gaining stream at time = 50 years. A line across the simulation domain denotes the stream. This simulation forces all of the contaminated water discharge to the stream because ground water hydraulic gradients on either side of the stream converges toward it. For the contaminant sources assumed in this study, the source duration ranges from a few years to several decades; accordingly, the simulations are made to assess the propensity of the plumes to be fully discharged to the stream within 100 years of time. It is assumed that during the simulation period, contaminated water in the aquifer will not be used for human consumption.

Under Type II gaining stream conditions, the discharge of contaminated water to a stream depends upon the difference (Δh_{stream}) between the water levels in the stream and in the aquifer, and on the stream bed conductance (C_{stream}). Figures 3(a) through 3(c) present the results of simulations assuming Δh_{stream}=1, 4, and 7 m, respectively, for the case of C_{stream} = 1000 m/yr. These figures clearly illustrate that, as the Δh_{stream} is increased, more contaminated water is captured by the stream. For Δh_{stream} = 7 m, all of the contaminated ground water flows into the stream.

Simulation of dilution by losing streams

The input parameters used in the gaining stream analyses are also adopted for simulations of dilution effects by losing streams. The total flow rate from a stream to an aquifer is the unit length loss rate (Q_{stream}) times the length of the stream. Two scenarios are simulated for unit length loss rates of 100, and 4000 m^2/yr, respectively, as shown in Figures 4a through 4b. Comparison of Figure 4a with Figure 2a (no stream case) demonstrates that for Q_{stream}=100 m^2/yr, the influence of stream losses on contaminant concentration is virtually negligible. The extent of the concentrations and the plume are reduced as the unit length loss rate is increased from 500 m^2/yr (Figure 4a) to 4000 m^2/yr (Figure 4b).

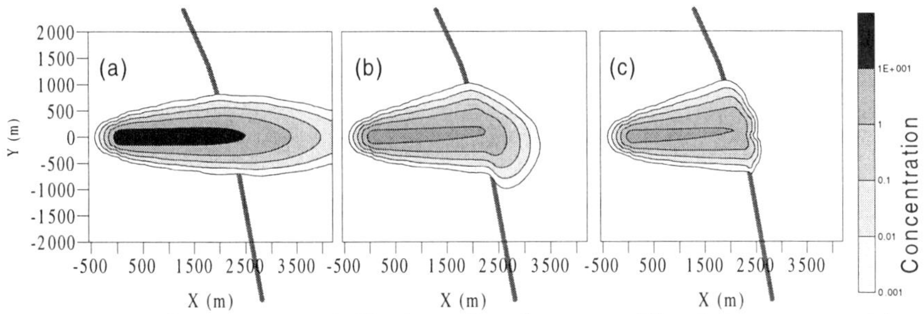

Figure 3. Simulations of dilution effect by a type II gaining stream. (a) Δh_{stream} = 1 m; (b) Δh_{stream} = 4 m; (c) Δh_{stream} = 7 m.

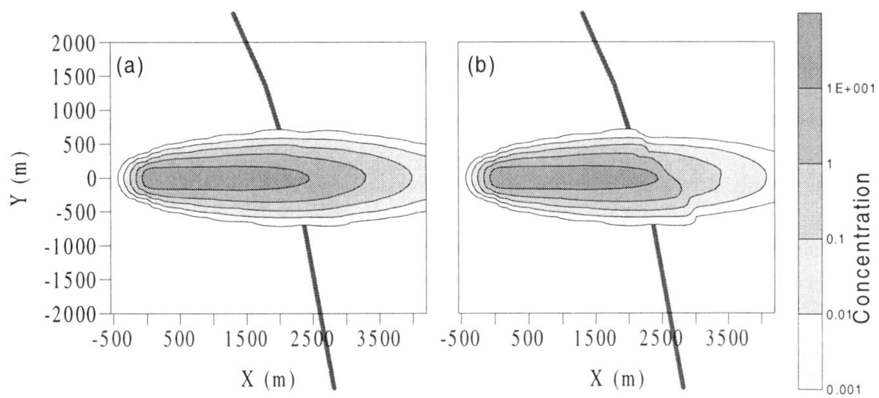

Figure 4. Simulations of dilution effect by a losing stream. (a) Q_{stream} = 100 m^2/yr; (b) Q_{stream} = 4000 m^2/yr

Simulation of dilution by segmented streams

For consistency purposes, ground water model input parameters used in the previous sections are also used in the set of simulations designed to assess the influence of a segmented stream. As shown in Figures 5a through 5c, the stream consists of two segments with the dotted line representing a losing segment and the solid line representing a gaining segment. Under this mixed form of stream-aquifer interaction, with recharge from the losing segment subsequently discharging water into the gaining segment, the ground water flow field is influenced by the stream. In the gaining segment of the stream, the hydraulic head difference between the stream and aquifer is assigned a value of 5 m in all simulations. Parameters varied in the simulations are unit length loss rate for the losing segment and the unit length conductance for the gaining segment. Figures 5a through 5c show simulation results for various combinations of Q_{stream} and C_{stream}. Clearly, as both the Q_{stream} and C_{stream} are simultaneously increased, the dilution effects created by the stream are much greater than that would occur in comparable simulations based solely on a losing or gaining stream. In addition, the deflection of the plume from its original transport path increases as both Q_{stream} and C_{stream} are increased.

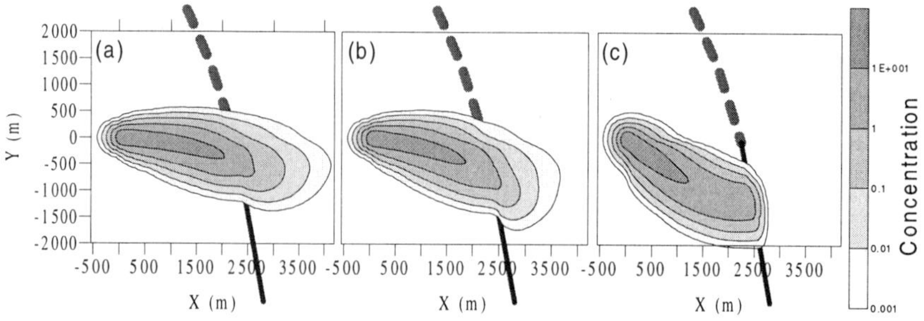

Figure 5. Simulation of dilution effects by a segmented stream. (a) Q_{stream} = 100 m^2/yr and C_{stream} = 100 m/yr; (b) Q_{stream} = 200 m^2/yr and C_{stream} = 200 m/yr; (c) Q_{stream} = 500 m^2/yr and C_{stream} = 500 m/yr; (d) Q_{stream} = 1000 m^2/yr and C_{stream} = 1000 m/yr.

Natural flushing at a DOE site

Application of the previously-discussed Type I gaining stream phenomena at a former uranium processing site falling under the preview of DOE's Uranium Mill Tailings Remedial Action (UMTRA) Project is made to illustrate the potential for total natural flushing of ground-water contamination. The site encompasses about 80 acres on a level, alluvial terrace between the Wind River and the Little Wind River near Riverton, Wyoming. The uranium mill was operated from 1958 to 1963. The tailings pile and associated contaminated soils were removed in 1988 and 1989. Ground water occurs in an unconfined surficial aquifer, an underlying semi-confined aquifer, and a deep confine7d aquifer. The surficial aquifer consisting of sand and gravel (5-6 m thick), and the semi-confined aquifer of sandstone (5-9 m thick), are separated by a layer of shale and siltstone (2-3 m thick). The Baseline Risk Assessment Report for this site indicates that the greatest concern for ground-water contamination is related to processes occurring in the surficial aquifer (DOE, 1995). Previous investigations have suggested that contaminants in the aquifer will be flushed from the aquifer by discharge to the Little Wind River (Knowlton et al, 1998). In this exercise, simulation is considered only for the surficial aquifer. Monte Carlo simulations are first made using site-specific flow and fate transport parameters. Simulations determined to reasonably match historical observations of contaminant concentration in monitoring wells through a post-conditioning exercise are then used to predict future distributions of the contaminants.

A total of 100 Monte Carlo realizations of uranium transport were simulated and conditioned using 16 concentration measurements at a conditioning time t_c= 39 years after the waste was released into the aquifer. From among the 100 simulations, 49 of them passed a criterion of RMSE = 0.35 mg/L between the calculated and measured concentrations with most of the RMSE values being close to zero.

Simulations meeting the conditioning criterion were continued for 100 years beyond t_c. Figures 6a through 6d show average uranium concentrations from these

49 simulations at 0, 50, 75, and 100 years beyond t_c, respectively. At simulation time = t_c+75 years, only traces of contaminant were left near the stream (Figure 6c), all of which were flushed into the stream as the simulation was continued for another 25 years (Figure 6d).

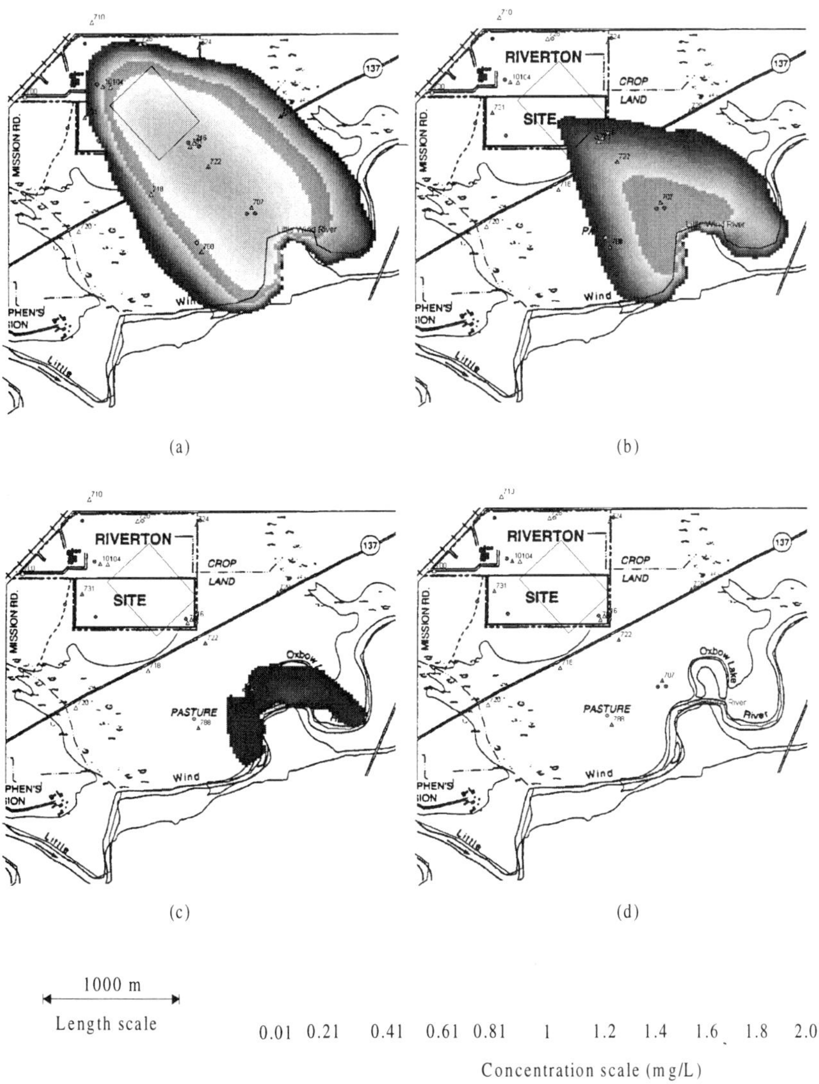

Figure 6. Average concentrations. (a) Time = t_c years; (b) Time = t_c+50 years; (c) Time = t_c+75 years; (d) Time = t_c+100 years.

CONCLUSIONS

This paper uses computer simulations to investigate contaminant attenuation created in ground water by area recharge and stream-aquifer interaction. The stream-aquifer processes considered include losing streams, gaining streams, and combined losing and gaining (segmented) streams. The simulations indicate that all of the investigated processes contribute to the natural attenuation of contaminant concentrations. Sensitivity analyses performed by varying recharge and stream parameters show the relative effects each has on the total dilution process. An example application at a DOE facility is made to demonstrate the potential of using natural flushing as a remedial alternative.

In this study, it was assumed that the flow rate in the stream was much larger than the flow rate from the aquifer to the stream so that contaminated ground water was significantly diluted by stream water. A well-designed contamination monitoring system may be necessary to determine the progress of the natural attenuation process. The methodology presented in this paper has been programmed into a Windows 95-based computer code (Knowlton et al, 1998).

Acknowledgement

Funding of this study was partly provided by Sandia National Laboratories under U.S. DOE contract DE-AC04-94AL85000. Contributions by Dan McCorQuodale, Janis White, and Douglas Walker to the development of the modeling software are appreciated.

REFERENCES

Chapelle, F.H., P.M. Bradley, D.R. Lovely, and, D. A. Vroblesky, 1996. Measuring rates of biodegradation in a contaminated aquifer using field and laboratory methods. Ground Water 34, no.4: 691-698.

U.S. Department of Energy, 1995, Baseline risk assessment of ground water contamination at the uranium mill tailings site near Riverton, Wyoming. DOE/AL/62350-65, Rev. 1 c.1. Available at NTIS.

Knowlton, R. G. Jr., D. M. Peterson, and H. Zhang, 1998, Analysis of natural attenuation as the preferred ground water remedial alternative at the DOE UMTRA site near Riverton, WY. Sandia National Laboratries, 93pp.

Metzler, D., R. Plieness, R. Knowlton, D. Peterson, H. Zhang, and J. Knott, 1997, Implementing the natural flushing strategy: A approach to meeting the EPA ground water standards at select uranium mill tailings sites, Tailings and Mine Waste 97, Balkema, Rotterdam, ISBN 90 54108576.

Rafai, H.S., P.B. Bedient, J. T. Wilson, K. M. Miller, and J. M. Armstrong, 1987, Biodegradation modeling at aviation fuel spill site. Journal of Environmental Engineering (American Society of Civil Engineers), 114, no.5: 1007-1029.

A PROBABILISTIC MODEL OF TIME-TO-CLEANUP BY NATURAL ATTENUATION

Paul C. Chrostowski, Damian V. Preziosi, Lorraine J. Pearsall, Sarah A. Foster, (CPF Associates, Inc., Takoma Park, MD, USA.)

ABSTRACT: When a natural attenuation remedy is being considered for a contaminated site, it is important to be able to calculate a reliable and accurate time-to-cleanup. This allows comparison to other remedial alternatives and allows risk managers to make scientifically defensible decisions based on data. All calculations of complex environmental phenomena involve uncertainty and variability. By accounting for the uncertainty and variability in calculations of time-to-cleanup, risk managers are able to put error bounds on calculated estimates and arrive at reliable conclusions. This paper explores the sources of uncertainty involved in calculating time-to-cleanup of halogenated solvents and illustrates the use of uncertainty analysis techniques in understanding degradation processes at a site.

INTRODUCTION

The feasibility of applying a particular remedy to a contaminated site often involves making a determination of the duration of remediation. If a natural attenuation remedy is being considered, many regulatory agencies (USEPA 1999) require an estimate of the time-to-cleanup that may be used as a point of comparison with active remedies. The calculation of time-to-cleanup is usually accomplished through the application of a mathematical model which relies on some form of the convection-dispersion equation that focuses on degradation processes, especially biodegradation. Estimates of environmental parameters calculated by mathematical models are subject to various forms of uncertainty, usually characterized as parameter variability, measurement uncertainty and modeling uncertainty. By explicitly considering and defining these sources of uncertainty, a model is able to put error bounds around the final estimates, thus giving risk managers a more reliable estimate of the time-to-cleanup. This paper explores the most significant forms of uncertainty involved in calculating time-to-cleanup in both the saturated and unsaturated zones. Methods are presented for accounting for uncertainty in numerical calculations and illustrated by application to data at an actual site.

UNCERTAINTY IN MATHEMATICAL ESTIMATES

All characterizations of physical reality that are based on mathematical models are inherently uncertain, as are all measurements of physical reality. A presentation of modeling results is incomplete without an assessment of the uncertainty in the results. Regarding the duration of remediation (or time-to-cleanup), a statement that remediation by natural attenuation will take 10 years compared to 7 years for an active remedy has a vastly different meaning than the statement that natural attenuation will take 10 ± 2 years compared to 7 ± 5 years for the active remedy. In the first instance, there appears to be a significant difference between the alternatives, while in the second instance, there may be no significant difference.

In discussing good modeling practice, Trapp and Matthies (1997) note that uncertainty of modeling results comes from the variance of the input data (stochastic error) as well as from inadequate modeling structures. Finkel (1990) also discusses measurement uncertainty and the error that stems from excluding variables. In general stochastic uncertainty is due to natural variation in a variable which cannot be reduced by further measurement, e.g., soil organic carbon content can vary widely throughout the unsaturated zone at a site. Measurement uncertainty can be reduced by improving the measurement. For example, the precision of chemical analyses can almost always be improved by enhanced analytical and extraction techniques. It is often difficult to separate measurement uncertainty from parameter variability. For example, without a detailed analysis, it is usually not possible to determine how much of the difference in pH measurements in groundwater underlying a site is due to actual pH variability or lack of precision in pH measurements. Although techniques exist for separating measurement uncertainty from variability (such as nested Monte Carlo and Grubb's test), they are beyond the scope of this paper and will not be discussed further here.

Model Uncertainty. There are several sources of model uncertainty, however, the most important is probably the selection of the incorrect form for the model. This can be illustrated by a digression into the chemical engineering kinetics of natural attenuation. The rate at which a chemical in the subsurface environment will attenuate to the point at which a performance goal is reached is determined by both chemical processes (biodegradation, chemical degradation) and physical processes (convection, volatilization). The well-known convection-dispersion equation (Bigger and Nielsen 1967) includes terms for hydrodynamic dispersion, convection, sorption, and chemical reaction. In very simple terms:

$$V\frac{dC}{dt} = F + Vr \tag{1}$$

Where V = reaction volume
dC/dt = rate of accumulation (degradation) in concentration
F = convective, dispersive, or advective flux, and
r = reaction rate

If we neglect transport processes, the problem of natural attenuation then focuses on proper characterization of the reaction rate.

In the environment, reactions are typically zero order, first order, second order, or Michaelis-Menten (enzymatic) in nature. Other orders including fraction orders may also occur based on the complexity of the process (see Cusack 1999 and Pilling and Seakins 1995 for a further discussion of chemical reaction kinetics). Thus, a first step in avoiding excess model uncertainty is the selection of an appropriate reaction order. Zero order reactions, exemplified by radioactive decay, do not depend on the concentration of the chemical undergoing degradation. First order reactions depend on the concentration of the chemical undergoing degradation. Second order reactions depend on the concentration of the chemical undergoing degradation in addition to the concentration of another chemical (e.g., ozone or hydroxyl radical). Hydrolysis reactions are significant second order environmental processes. Michaelis-Menten (or Monod) kinetics depends on characteristics of an enzyme or other complex biological system. In many cases, second order or Michaelis-Menten kinetics reduce to pseudo first order reactions without loss of mathematical accuracy. The outcome of reaction order analysis is the selection of the appropriate form for a reaction rate constant which is used further in the analysis.

Although the reaction rate term (r) typically excludes physical processes, some, such as volatilization are amenable to treatment in a pseudo-first order fashion. This is an important consideration for chemicals such as volatile organic solvents where vapor phase transport is important. Movement in the vapor phase is typically described by integration of Fick's law of diffusion:

$$F_v = D\left(C_i - C_{eq}\right) \tag{2}$$

Where F_v = volatilization flux
D = medium-specific diffusivity
C_i = instantaneous concentration, and
C_{eq} = equilibrium concentration.

If we substitute a mass transfer coefficient and specific interfacial area for diffusivity and assume that $C_{eq}>>C_i$, then $C_i = 0$ and this takes the functional form of a first order rate process:

$$\frac{dC}{dt} = -\frac{K}{a}C = -k_v C \tag{3}$$

Where K = mass transfer coefficient
a = specific interfacial area, and
k_v = first order volatilization rate constant.

Uncertainty associated with selection of the wrong reaction order often happens with respect to hydrolysis. Hydrolysis occurs through different mechanisms of action in basic, acidic, and neutral environments. Oftentimes, the presence of acid or base catalyzes the reaction and accelerates its rate. An assumption that the reaction is first order may tend to under-estimate the rate of the reaction if the contribution of acid or base is not taken into account.

Once the order of the reaction has been determined, an appropriate reactor model must be selected. Two useful models for environmental processes are the continuously mixed flow reactor and the plug-flow reactor. The continuously mixed reaction conceptually fits the situation of a relatively immobile chemical compound that often remains in a single unsaturated zone location where it is subject to degradation processes. For example, a spill of benzo(a)pyrene will remain near the soil surface due to a large sorption coefficient. While at the surface, it may be subject to chemical processes such as biodegradation and physical processes such as volatilization. The time to cleanup for this type of process is given by:

$$t = \frac{1}{k}\left[\frac{C_t}{C_o} - 1\right] \tag{4}$$

Where C_o = initial chemical concentration,
C_t = target chemical concentration, and
k = any or all first order rate constants.

The plug flow reactor is useful in describing the behavior of a compound that is being transported either horizontally through the unsaturated zone or laterally in groundwater. A good example would be a spill of trichloroethylene on the surface which would infiltrate with rain water and be subject to biodegradation and volatilization during the period of transport. The time-to-cleanup for a plug flow model is given by:

$$t = -\ln\left(\frac{C_t}{C_o}\right)\frac{1}{k} \tag{5}$$

The difference in the outcomes of the two models can be shown by a simple calculation. If we assume that Co = 100 μg/L and Ct = 5 μg/L with a first order rate constant of 0.1 $year^{-1}$, the time to cleanup assuming a plug flow model is 5.3 years whereas the time to cleanup assuming a continuously mixed model is 9.5 years.

Parameter Variability. In modeling time-to-cleanup through natural attenuation, parameter variability is often reflected in calculation of a rate constant. The best way to illustrate this is through the examination of a hypothetical degradation process. This process assumes first order kinetics for the disappearance of a chemical from the subsurface. This particular chemical may be biodegraded, hydrolyzed, or volatilized. The overall first order rate constant is given as the sum of the individual rate constants:

$$k_o = k_b + k_h + k_v \tag{6}$$

Where k_o = overall first order rate constant
k_b = biodegradation rate constant
k_h = hydrolysis rate constant, and

k_v = volatilization rate constant.

In this case, we assume that hydrolysis is base catalyzed, thus k_h is a pseudo-first order rate constant defined by:

$$k_h = k_2[OH^-] \quad (7)$$

Where k_2 = second order rate constant for base-catalyzed hydrolysis, and [OH$^-$] = hydroxide ion concentration.

Further, we assume that volatilization may be calculated by the relation of Bomberger et al. (1982) as a function of a series of constants and the depth over which the chemical volatilizes:

$$k_v = \frac{K'}{Z^2} \quad (8)$$

Where K' = constant incorporating mass transfer functions, and Z = volatilization depth or path length.

Thus, there are three explicit parameters that may exhibit variability – the biodegradation rate constant, the hydroxide ion concentration, and the volatilization depth. In reality, the mass transfer term (K') incorporates physical properties of the environment such as total porosity, air-filled porosity, water-filled porosity, organic carbon content, and bulk density, however, for the sake of simplicity in this discussion (and because depth is more significant to the outcome), these are being treated as a constant.

Rate constants for biodegradation are available from the scientific literature. They may input to a kinetic model as probability distributions to be able to describe the variability associated with biodegradation. The variability in hydrolysis rate constants is usually associated with variability in pH. If a number of pH measurements are made at a site, they may be input to a kinetic model as a probability distribution after transformation to [OH$^-$]. The variability in depth can be accounted for as a probability distribution based on site observations. Other parameters that are often encountered in time-to-cleanup models such as oxidant concentration or site characteristics such as porosity and organic carbon content can also be input as probability distributions. Variables that change in response to environmental phenomena (such as diurnal change of temperature or change in water-filled porosity after precipitation) are best modeled as dynamic events in a sensitivity analysis mode.

Once the probability distributions and modeling regime have been decided upon, they need to be combined to produce an estimate of overall uncertainty in the results. There are numerous techniques for propagating uncertainty that have been traditionally used in groundwater modeling including Monte Carlo simulation, second moment approximation, and Bayesian analysis (Roberds 1990, Smith et al. 1990). Each of these techniques has assets and drawbacks. A particular technique should only be selected after careful evaluation of the chemical and physical situation as well as a thorough understanding of the statistical theory providing the basis for each technique.

ILLUSTRATION

The techniques used above were used to calculate the time-to-cleanup for trichloroethylene (TCE) in groundwater at a contaminated site. A plug flow model was chosen for the analysis. Equation (6) was parameterized with a combination of site-specific and literature values. A probability distribution was obtained for the biodegradation rate constant based on data provided in Syracuse Research Corporation's database for anaerobic degradation of organic chemicals in groundwater (SRC 1999). TCE biodegradation rates were found to be log-normally distributed with a geometric mean of 2.5 X 10^{-3} day^{-1} and a geometric standard deviation of 4.4 day^{-1}. The hydrolysis rate for TCE was found to not depend on pH (Mackay et al. 1993) and was input as a constant. The value for K' was calculated from site specific considerations to be 0.05 cm^2 day^{-1}, and Z was input as a uniform distribution with a range from 45,000 cm to 180,000 cm. A Monte Carlo simulation (Latin hypercube mode) and associated sensitivity analysis were conducted on the overall rate constant (k_0). After 10,000 iterations, the value for ko was found to range from 0.0002 day^{-1} to 0.09 day^{-1} with a mean of 0.01 day^{-1} and a standard deviation of 0.01 day^{-1}. The results of an accompanying sensitivity analysis revealed that both hydrolysis and volatilization were insignificant compared to the biodegradation of TCE under these circumstances.

It is instructive to compare these results to monitoring data that were obtained at the site. Monitoring data for TCE concentrations were obtained in a series of wells over a period of about 2 years. The data were fit to a series of hypothetical curves. Since the best fit was obtained with a plot of ln(C/Co) against time, it was decided that degradation was a first order plug flow process. The mean k_0 calculated from these data was 0.009 day^{-1} with a standard deviation of 0.05 day^{-1} which are clearly in the range established by the model.

The correspondence between the modeling and the measured results shows that biodegradation dominates the calculation of the overall rate constant at this site rather than hydrolysis, vapor phase transport or dilution. This can be important in a regulatory context since many regulatory agencies require a showing that chemicals are actually degrading. The correspondence between the measurements and the model when taken with other information such as the occurrence of daughter products is strong evidence that TCE is actually degrading. The results from the model were also used to calculate time-to-cleanup for the site. The average time to-cleanup was about 1.5 years with a 95% confidence limit from 0.7 years to 3.4 years. Since engineering calculations revealed that the site could be remediated with groundwater extraction followed by air stripping in two years there was no time advantage to implementing an active remedy.

CONCLUSIONS AND FUTURE WORK

The work that we have conducted to date shows that uncertainty associated with parameter variability and model selection can be defined by application of some simple techniques. Better definition of uncertainty in time-to-cleanup can lead to more reliable risk management decisions and can focus future research and investigations on the more uncertain variables. Currently, we are expanding this work to account for a fuller range of environmental and chemodynamic factors, broadening the time calculations to include active as well as passive remedies, and developing simple statistical applications to resolve parameter variability from measurement uncertainty.

REFERENCES

Biggar, J.W. and Nielsen, D.R. 1967. Miscible displacement and leaching phenomena. *Agronomy* 11:254-274.

Bomberger, D.C., Gwinn, J.L., Mabey, W.R., Tuse, D., and Chou, T.W. 1982. "Environmental fate and transport at the terrestrial-atmospheric interface". In Swann and Eschenroeder Eds. *Fate of Chemicals in the Environment.* AS. Washington, DC.

Cusack, R.W. 1999. "A fresh look at reaction engineering". *Chemical Engineering*, October pp. 134-146 and December pp 80-85.

Finkel, A.M. 1990. *Confronting Uncertainty in Risk Management.* Resources for the Future, Washington, DC.

Mackay, D., Shiu, W.Y. and Ma, K.C. 1993. *Illustrated Handbook of Physical-Chemical Properties and Environmental Fate of Organic Chemicals.* Volume III. Lewis Publishers, Boca Raton, FL.

Pilling, M.J. and Seakins, P.W. 1995. *Reaction Kinetics.* Oxford Science Publications. New York.

Roberds, W.J. 1990. "Risk-based decision making regarding mixed-waste systems". In McTernan and Kaplan, eds. *Risk Assessment for Groundwater Pollution Control.* ASCE, New York pp187-268.

Smith, L., Freeze, R.A. and Massmann, J. 1990. "Geostatistical approach to site characterization and risk assessment related to groundwater contamination at hazardous waste management sites". In McTernan and Kaplan, *op. cit.* pp. 281-311.

Syracuse Research Corporation (SRC) 1999. http://esc-plaza.syrres.com/resrchpr.htm.

Trapp, S. and Matthies, M. 1997. *Chemodynamics and Environmental Modeling*. Springer, New York.

U.S. Environmental Protection Agency (USEPA) 1999. *Use of Monitored Natural Attenuation at Superfund, RCRA Corrective Action, and Underground Storage Tank Sites*. OSWER Directive No. 9200.4-17P.

CHANGES IN THE NATURAL ATTENUATION OF PERCHLOROETHYLENE ASSOCIATED WITH SPARGING

Mark Ferrey, Jim Lundy, and Paul Estuesta. Minnesota Pollution Control Agency, St. Paul, Minnesota, USA

Abstract: Spills from a drycleaner solvent distributor center in Minnesota caused perchloroethylene (PCE) contamination of a surficial aquifer. Ground water remediation efforts included an *in situ* ground water circulation system ("vacuum vaporizer wells") designed to remove volatile contaminant mass by in-well air sparging. Ground water data show that prior to sparging in 1994, the biodegradation rate for PCE was -2.4 yr^{-1}. Following the startup of the vacuum vaporizer well, concentrations of the breakdown products of PCE – trichloroethylene (TCE), *cis*-1,2-dichloroethylene (cDCE), and vinyl chloride (VC) – decreased near the source area but increased in ground water downgradient. Concentrations of all these contaminants returned to pre-sparging concentrations with the deactivation of the vacuum vaporizer well in 1997. The overall size of the plume did not appear to increase over the ten-year study period. This suggests that oxygenation of the aquifer by the vacuum vaporizer well interfered with the reductive dehalogenation of contaminants that was occurring naturally in ground water. The subsequent delay of the anaerobic degradation process caused increased breakdown product concentrations in downgradient portions of the plume.

INTRODUCTION

Recalcitrant chlorinated compounds such as PCE or TCE can biologically degrade under highly reducing ground water conditions. This understanding has changed the approach to remediating sites contaminated with these compounds. Natural attenuation may constitute the sole remedy for contaminated ground water when evaluated within the context of a particular site (Environmental Protection Agency, 1998).

The most effective remedial actions may be those that augment the natural attenuation capacity of a site (Chapelle and Bradley, 1999). When "active" remediation systems are selected, it is important to consider the effect they will have on the ground water geochemistry and intrinsic biodegradation processes at the site (Norris et al., 1999). Technologies that oxidize the subsurface may disrupt conditions necessary for biological reductive dehalogenation. At the site discussed here, in-well air sparging was applied to remediate an aquifer contaminated by PCE. The aeration of ground water resulting from the remedial approach may have caused subsequent changes in the distribution of contaminant breakdown products with distance from the source.

SITE DESCRIPTION AND METHODS

Geology and Hydrogeology. The hydrogeologic unit of interest (upper sand) consists of coarse to fine glacial outwash sand to a depth of 30 feet, underlain by fine sand (30-75 feet) and bedrock. The upper sand is continuous across the site. Ground water occurs at a depth of 12 feet and flows east or southeast under a hydraulic gradient of approximately 0.005. Effective porosity is assumed to be 0.30. Aquifer pumping test data indicate a horizontal hydraulic conductivity of 0.046 cm sec^{-1} (130 ft day^{-1}), yielding a ground water velocity of 790 ft yr^{-1}. The fraction of organic carbon (f_{oc}) in these sediments is 0.02%.

PCE released from a former dry cleaner supply company contaminated the upper sand aquifer. Site investigations identified a contaminant plume of PCE, TCE, cDCE, and VC approximately 800 feet long and 200 feet wide. No evidence of dense, non-aqueous phase liquid (DNAPL) was found. Figure 1 is a site map showing selected monitoring well locations, ground water elevations, and approximate plume boundaries.

Field Methods. Monitoring wells are two inches in diameter, with 10-foot long stainless steel screens straddling the water table. Before sampling, wells were purged until dissolved oxygen (DO), redox potential (ORP), specific conductance, pH, and temperature readings were stable using probes in a low-flow cell apparatus (YSI Inc., Yellow Springs, OH).

Additional analyses included manganese II, iron II, nitrate, sulfate, sulfide, chloride, ammonia, methane, dissolved organic carbon (DOC), and ethene. Field test kits (Hach Co., Loveland, CO) were used for the manganese and iron analyses. Nitrate, sulfate, sulfide, chloride, ammonia, methane, dissolved organic carbon (DOC), volatile organic compounds (VOC), and ethene were analyzed using standard procedures at a fixed-base laboratory.

Vacuum Vaporizer Well. Vacuum vaporizer wells have two screens that bracket the plume; one straddles the water table and one is placed deeper in the aquifer. Casing separates the two screens. Ground water is stripped of VOC by an in-well sparger, and the cleaned (and oxygenated) water is returned to the aquifer through the upper well screen by means of a pump, supporting ground water circulation within the aquifer. Bentonite separates the two screened sections to block short-circuit flow. During operation, the air pressure drop in the casing provides remedial benefit similar to that of a soil vapor extraction (SVE) system. The expected ground water circulation cell radius depends upon screen lengths and aquifer properties such as aquifer thickness, ground water flow velocity, and aquifer anisotropy.

From September 1994 to August 1997, a vacuum vaporizer well was operated to remove VOC mass from the source area (Lundy, 1995). The screens were placed across the water table at 12-20 feet and at 25-29 feet, both within the upper sand.

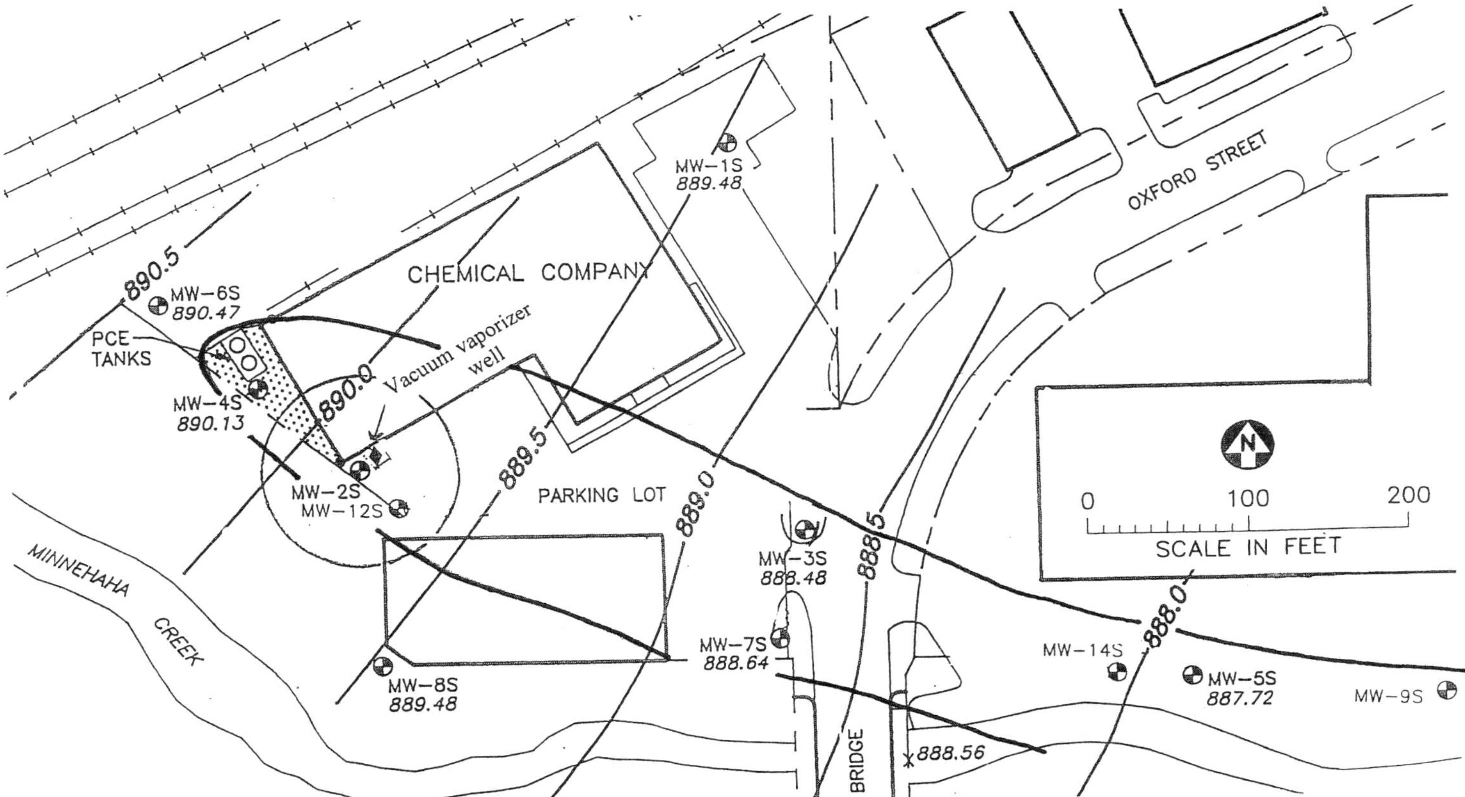

Figure 1. Site map showing the locations of monitoring wells, approximate PCE plume boundary, and the vacuum vaporizer well installed in 1994. Wells 6S, 4S, 2S, 7S, 14S, 5S and 9S lay in the longitudinal axis of ground water flow, along the plume centerline. Known PCE spills occurred in the area between the PCE tanks and MW-2S. (Modified from Barr Engineering, 1992.)

RESULTS

Contaminant concentration trends. Figure 2a shows PCE concentrations along the ground water flowpath before, during, and after the vacuum vaporizer well operated. The highest PCE concentrations were observed in well 4S, nearest the source area, before installation of the vacuum vaporizer well. PCE levels decreased from an average of 476 μg/L to less than 2 μg/L in well 4S near the source area after six months of operation. However, PCE concentrations increased 360 feet downgradient (approximately 0.5 years advective flow time) in well 7S from 54 to 150 μg/L.

Similarly, TCE and cDCE concentrations were highest near the source area prior to ground water sparging and decreased in well 4S over the time that the remediation system operated. However, time-averaged TCE, cDCE and VC concentrations (Figs. 2b-d) increased from 6 to 20 μg/L, 82 to 180 μg/L, and 3 to 25 μg/L, respectively, 360 feet downgradient (well 7S) during vacuum vaporizer well operation. Similar trends were observed for total VOC in well 7S, where concentrations increased from 144 to 358 μg/L during operation of the vacuum vaporizer well (data not shown).

Following vacuum vaporizer well deactivation, the highest TCE and cDCE concentrations were again observed near the source area, rebounding from 3 to 13 μg/L and 6 to 44 μg/L, respectively (Fig. 2b,c). Downgradient, at well 7S, they decreased to 4 and 29 μg/L, respectively. VC concentrations also decreased at this time in well 7S from 25 to 3 μg/L (Fig. 2d). PCE concentrations remained low in the source area well 4S following the termination of sparging. Total VOC concentrations decreased at 7S to an average of 69 μg/L after system deactivation. Concentrations of all contaminants remained low in monitoring wells located 640 feet to 775 feet from the source area (wells 5S and 9S, respectively) over the entire nine year time period of this study.

Pre-remediation attenuation kinetics. PCE concentrations in ground water from wells placed along the suspected axis of the contaminant flowpath (4S, 2S, 7S, 14S, and 5S) are plotted as a function of distance from the source in Figure 3, using 1994 data.

A site-specific estimate of the attenuation rate requires the assumption that the plume is stable (i.e., not expanding). The regression in Figure 3 is in units of k_{PCE} / v_x, where k_{PCE} is the overall attenuation rate for PCE, and v_x is the ground water velocity. With v_x estimated at 790 feet year^{-1} from a pump test, k_{PCE} was -4.8 yr^{-1} (upper 95% C.I. = -3.4 yr^{-1}) prior to vacuum vaporizer well activation in September, 1994.

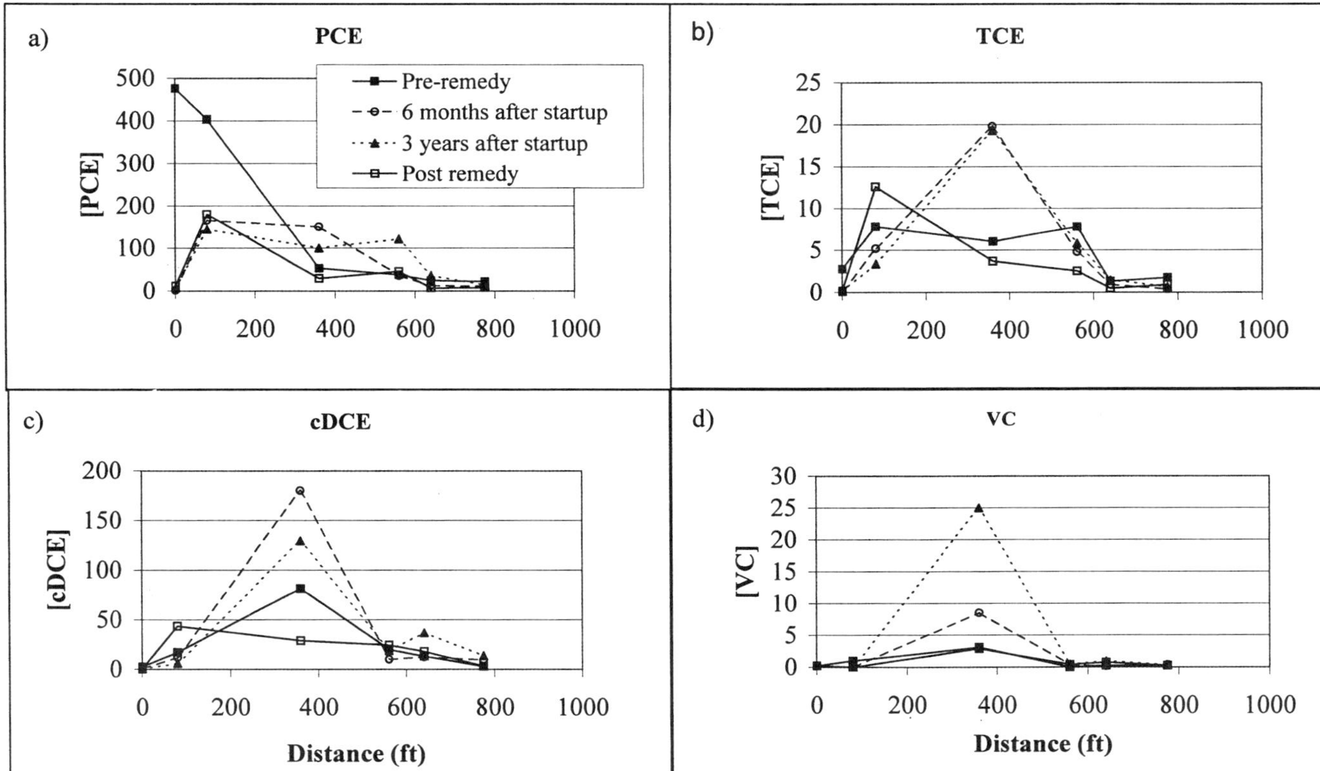

Figures 2a-d. PCE, TCE, cDCE, and VC concentrations vs. distance along the ground water flow path before, during, and after the vacuum vaporizer well startup. Data points represent the average of four consecutive sampling events. Concentrations are in ug/L. (Note variations in the vertical scale.)

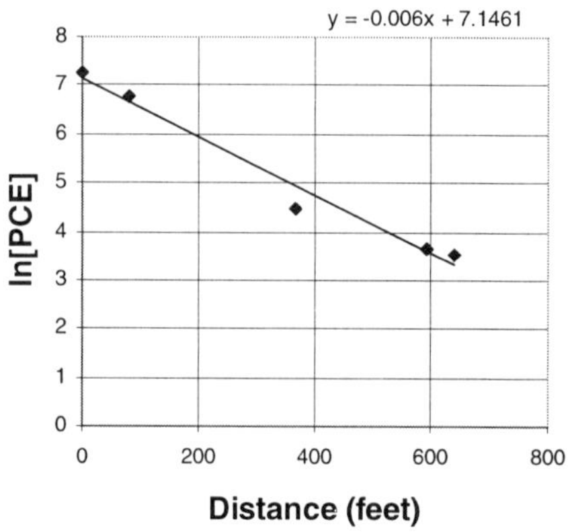

Figure 3. PCE concentration (µg/L) versus distance prior to operation of the vacuum vaporizer well in 1994.

We estimate the biodegradation rate, λ, according to the method of Buscheck and Alcantar (1995):

$$\lambda = \left(\frac{v_c}{4\alpha}\right)\left[\left(1+(2)(\alpha)\left(\frac{k}{v_x}\right)\right)^2 - 1\right]$$

where:

v_c = contaminant transport velocity in ground water, estimated by v_x/R_f,;

α = scale dependent dispersivity, usually estimated at 5 to 10% of plume length (at our site, 30 to 65 ft.) Assuming α=32 feet, λ_{PCE} is approximately -3.55 yr^{-1}. Using the upper 95% confidence interval for k_{PCE} (– 3.4 yr^{-1}), we estimate λ_{PCE} at – 2.43 yr^{-1}, or 71% of k_{PCE}.

DISCUSSION AND CONCLUSIONS

Our data show PCE was attenuating in ground water at a high rate prior to the remedial efforts. This led us to re-evaluate the overall benefit of the vacuum vaporizer well compared to intrinsic reductive dehalogenation in ground water.

Ground water geochemistry (Table 1) supports our hypothesis that operating the vacuum vaporizer well unintentionally disrupted *in situ* reductive dehalogenation of chlorinated contaminants by oxygenating the nearby ground water. Geochemical data was not collected prior to system startup. However dissolved oxygen concentrations within the designed radius of influence were approximately 3 mg/L (wells 12S and 2S) during system operation. Oxygen levels declined in the two months following system shutdown, indicating the vacuum vaporizer well aerated ground water to a downgradient distance of 30 feet.

Table 1. Ground water geochemistry during and after vacuum vaporizer well operation. All concentrations are in mg/L. Well 6S is a background well.

Well	Dist. (ft)	Date	O_2	NO_3^{-2}	Mn^{+2}	Fe^{+2}	H_2S	CH_4
6S	-	Apr., 1997	0.31		0.65	3.2		
		Oct., 1997	0.15	<0.05	0.5	2.0	<0.1	
2S	80	Apr., 1997	2.93		0.1	<0.05		
		Mar. 1998	1.18	0.34		0.4	<0.1	
12S	150	Apr., 1997	3.3	0.3			<0.1	<0.001
		Oct., 1997	0.3	0.1	0.2	<0.05	<0.1	0.001
7S	360	Apr., 1997	0.6	<0.05	<0.1	3.0	4.1	<0.001
		Oct., 1997	0.1	<0.05	0.7	1.5	<0.1	3.0

We believe that the increase in ground water redox conditions disrupted the anaerobic microbial reductive dehalogenation of PCE. This would have delayed reductive dehalogenation of PCE (and of TCE and cDCE) in the source area, allowing the metabolic breakdown products to migrate at higher concentrations downgradient to wells 7S and 3S. This is consistent with our observations of elevated TCE, cDCE, and VC concentrations in downgradient monitoring wells. Upon deactivation of the vacuum vaporizer well in September, 1997, the profile of contaminant concentrations versus distance reverted to the pre-remedial pattern of 1994.

VOC measurements in the vacuum vaporizer well's air effluent indicate that roughly 73 Kg of VOC were removed from the source vadose soil and ground water over the 978 days that the system operated. At peak removal rates, the system extracted 0.3 Kg day^{-1} over a 156 day period. Records indicate that at least 450 Kg of PCE was released at the site. Therefore, a maximum of roughly 16 percent of the released product was removed with the vacuum vaporizer well.

Assuming our lower estimate for the biodegradation rate for PCE of –2.4 yr^{-1} ($6.6x10^{-3}$ day^{-1}), it is possible that biodegradation initially eliminated up to 3 Kg day^{-1} PCE from the ground water prior to the sparging. Thus, it is possible that intrinsic biodegradation eliminated a much higher mass of contaminants over time than the vacuum vaporizer well.

This may explain why the contaminant mass removal effort appears to return so little benefit. We mathematically modeled the centerline of the plume with a one-dimensional analytic model, using input data for PCE retardation (1.6), ground water velocity (790 ft yr^{-1}), contaminant scale-dependent dispersivity (32 ft), an estimate of 10 ft for the contaminant source thickness, and a biological decay rate of 3.5 yr^{-1}. These simulations suggested that concentrations of contaminants found in the ground water in 1997 (after the SVE/sparging) might have been achieved solely through natural attenuation processes from the time the spills occurred in 1989, when source area PCE concentrations were about 10,000 μg/L. Any benefit to removing 16 percent of the contaminant mass through SVE and ground water sparging was difficult to discern through modeling, suggesting that it had only a marginal effect on the long term fate of this particular ground water plume.

Understanding the ground water geochemistry, the conditions that favor the intrinsic biodegradation of specific contaminants, and the rate of contaminant attenuation are important factors in evaluating the need for and design of remedial systems. Including these in initial site investigations may lead to an approach that optimizes existing conditions that favor the attenuation of ground water contaminants, and to more cost-effective remediation decisions.

REFERENCES

Barr Engineering Company. 1992. Phase II Remedial Investigation Report, Schloff Chemical and Supply Company Site, St. Louis Park, MN.

Buscheck, T. E., and C. M. Alcantar. 1995. "Regression techniques and analytical solutions to demonstrate intrinsic bioremediation." In R. E. Hinchee, J. T. Wilson, and D. C. Downey, (Eds.), *Intrinsic Bioremediation*, pp. 109-116. Battelle Press, Columbus, OH.

Chapelle, F. H., and P. M. Bradley. 1999. "Selecting remediation goals by assessing the natural attenuation capacity of groundwater systems." *Bioremed. J.* 2:227-238.

Environmental Protection Agency. 1998. *Technical Protocol for Evaluating Natural Attenuation of Chlorinated Solvents in Ground Water.* EPA/600/R-98/128.

Lundy, J. 1995. "The use of vacuum vaporizer wells to restore ground water at a perchloroethylene site in Minnesota.." In *Proceedings of the Water Resources Risk Conference*, American Institute of Hydrology, May 14-18, 1995.

Norris, R. D., D. J. Wilson, D. E. Ellis, and R. Siegrist. 1999. "Consideration of the effects of remediation technologies on natural attenuation" In B. C. Alleman, and A. Leeson (Eds.*), Natural Attenuation of Chlorinated Solvents, Petroleum Hydrocarbons, and Other Organic Compound,* pp. 59-64. Battelle Press, Columbus, OH.

EVALUATING THE POTENTIAL FOR CHLORINATED SOLVENT DEGRADATION FROM HYDROGEN CONCENTRATIONS

Mark Ferrey, Minnesota Pollution Control Agency, St. Paul, Minnesota, USA.
John Wilson and Don Kampbell, U.S. Environmental Protection Agency, Office of Research and Development, National Risk Management Research Laboratory, Subsurface Protection and Remediation Division, Ada, Oklahoma, USA.

ABSTRACT: Long-term monitoring of a large trichloroethylene (TCE) and 1,1,1-trichloroethane (TCA) ground water plume in Minnesota indicated that these contaminants attenuated with distance from the source. Mathematical modelling indicated that sufficient time had passed for the plume to fully develop. However, the actual concentrations of contaminants in the ground water were much lower than would be expected with dilution and dispersion alone, indicating that some other process was actively removing TCE and TCA along the flowpath. Although the presence of *cis*-1,2-dichloroethylene (cDCE) in the ground water was evidence of reductive dehalogenation, the absence of vinyl chloride (VC) and ethene suggested that the dechlorination process was incomplete. However, the concentrations of cDCE and other reduced products of degradation amounted to a small fraction of the missing TCE or TCA. Hydrogen concentrations exceeded 1.0 nM in ground water at many locations in the plume and at background locations, indicating that conditions are favorable for reductive transformation of chlorinated solvents throughout the entire plume. This suggests that a process other than reductive dehalogenation is removing the cDCE and preventing its accumulation along the ground water flowpath.

INTRODUCTION

Studies have shown that the most important biological process capable of transforming the chlorinated aliphatic compounds TCE and TCA is reductive dechlorination. Under highly reducing conditions, TCE can break down to form cDCE, VC, and ethene (Vogel and McCarty, 1985)

The geochemical environment that most favors reductive dehalogenation is one of sulfate reducing and methanogenic conditions. Under these conditions, microbial consortia can utilize chlorinated aliphatics as an electron acceptor. In addition, it has been shown that VC can biologically degrade under aerobic or iron reducing conditions, and that cDCE can anaerobically oxidize under manganese reducing conditions (Bradley et al., 1998).

Hydrogen is generated by microbial fermentation reactions in ground water, and its concentration in ground water often corresponds to the geochemistry of the aquifer (Lovley et al., 1994). Measuring the hydrogen concentration in ground water can help determine the terminal electron accepting process in the ground water and, in turn, help in an assessment of the potential for the reductive dehalogenation of chlorinated contaminants. The Technical

Protocol for Evaluating Natural Attenuation of Chlorinated Solvents in Ground Water (EPA, 1998) predicts that reductive dechlorination of solvents will occur if the concentration of hydrogen in the ground water exceeds 1 nanomolar (nM).

At a site in Minnesota, modelling shows that TCE and TCA concentrations are attenuating with distance along the ground water flowpath, and the presence of cDCE in the ground water is evidence that reductive dehalogenation is occurring although the geochemistry of the ground water is not considered favorable to the reductive dehalogenation process. However, hydrogen concentrations consistently exceed 1 nM at many locations in the ground water. These results may indicate an unusual pattern of natural attenuation of these contaminants.

SITE DESCRIPTION AND METHODS

The Twin Cities Army Ammunition Plant (TCAAP) is located in Minnesota north of St. Paul/Minneapolis. The disposal of waste solvents at the site resulted in a ground water plume of TCE and TCA in the deep ground water aquifer over five miles long.

There are four dominant hydrogeologic units at TCAAP. Unit 1 consists of glacial alluvium and lacustrine deposits of fine to medium grained sands and clayey silts, ranging in thickness from zero to 50 feet. Unit 2 comprises glacial till and sand and gravel lenses. Where present, Unit 2 is an aquitard to the vertical migration of ground water. Both Units 1 and 2 are discontinuous. Unit 3 consists of glacial outwash sands, varying in thickness from 25 to 450 feet, and Unit 4 is a bedrock formation of dolomite (Prairie du Chien Group) overlying quartz sandstone (Jordan Formation). The deep ground water plume is contained in Units 3 and 4.

Four rounds of ground water samples were collected over a two year period at six month intervals from June, 1996 through June, 1998 from monitoring wells located along what is assumed to be the central axis of the contaminant plume (Figure 1). Wells were purged until Eh readings, conductivity, and oxygen concentrations were stable. Oxygen, conductivity, pH, and Eh measurements were obtained through the use of a low flow cell (YSI Inc., Yellow Springs, OH). Reduced iron, reduced manganese, and sulfide were measured using field test kits (Hach Co., Loveland, CO); sulfate, chloride, nitrate, ethene, ethane, methane, and contaminants were measured in the laboratory using R. S. Kerr Research Center standard operating procedures.

Hydrogen. Ground water samples for the analysis of hydrogen were taken only from PVC wells using a pneumatically driven pump according to current EPA guidance (EPA, 1998). Samples were analyzed on location. A glass sampling bulb, fitted with a gas-tight septum, was nearly filled with ground water, allowing approximately 50 ml of air to remain in the bulb. Ground water was introduced to the bulb at approximately 400 ml/minute and allowed to escape from the bulb through a second port. Sequential sampling indicated that equilibration between the aqueous phase and the gaseous phase was complete at 30 minutes.

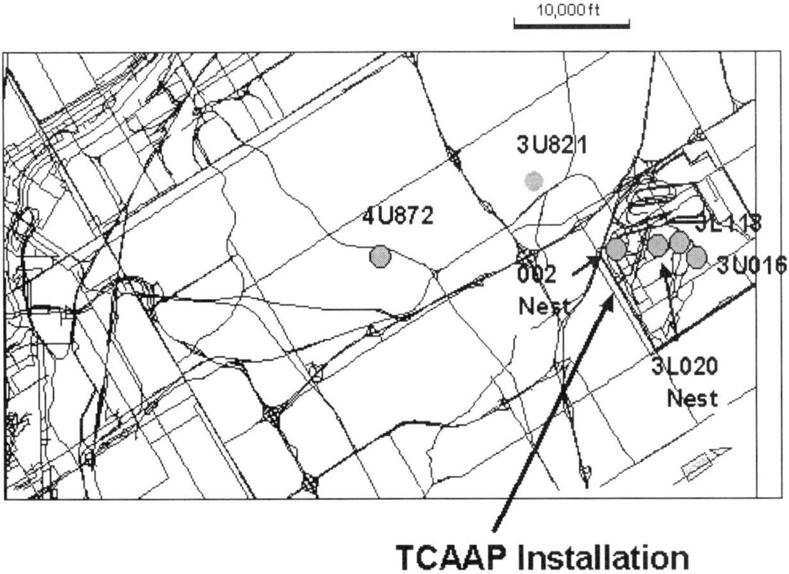

Figure 1. The TCAAP installation and monitoring wells sampled for this study. Wells 3L113 and 3U016 are background wells. The 020 and 002 well nests comprise the upper, middle, and lower portions of the Unit 3 aquifer.

Following the equilibration, a 10 ml aliquot of gas was immediately withdrawn from the bulb through the septum using a gas tight syringe and analyzed on a field chromatograph that was calibrated using hydrogen standards.

RESULTS AND DISCUSSION

The ground water geochemistry is shown in Table 1. Iron and manganese reducing conditions prevail at most locations, with no evidence of methanogenic or sulfate reducing conditions. Dissolved organic carbon concentrations were low, and no BTEX compounds (anthropogenic carbon) were found at any of the locations.

The presence of cDCE indicated that TCE was being biologically degraded through reductive dehalogenation. However, the absence of VC or ethene (Table 2) suggested that reductive dehalogenation was incomplete. The presence of 1,1-dichloroethylene (1,1-DCE) may be due to the abiotic breakdown of TCA.

EPA guidance predicts that under the geochemical conditions shown in Table 1, reductive dehalogenation processes are "occurring too slowly to contribute in a meaningful way" to the attenuation of the plume (U.S. EPA, 1998).

Hydrogen concentrations are usually in the range of 0.2 to 0.8 nM under iron reducing conditions (Lovley, 1994; EPA, 1998). Much of the ground water in the TCAAP plume contained hydrogen in this concentration range. However, at many locations, hydrogen concentrations were found along the axis of the

Table 1. Geochemistry of TCAAP ground water. Wells U016 and L113 are background locations. Concentrations are in mg/L unless otherwise noted.

Well	Unit	0_2	$N0_3^{-2}$	Mn^{+2}	Fe^{+2}	SO_4^{-2}	Methane	Eh (mv)	H_2 (nM)
U020	3	7.7	2.44	0.05	<0.05	26.4	<0.001	126	0.28
M020	3	0.7	<.05	1.12	0.05	2.59	0.054	104	1.41
L020	3	0.2	<.05	0.7	<0.05	4.02	0.023	-133	0.79
U002	3	1.4	0.48	0.07	<0.05	11.8	0.005	94	0.76
M002	3	0.2	0.5	1.2	<0.05	13.7	0.001	97	0.38
L002	3	0.1	<.05	1.4	<0.05	6.61	0.014	45	1.07
U821	4	0.5	0.29	0.06	0.7	12.3	0.001	-78	1.16
U872	4	0.3	<0.1	0.5	0.8	21.2	0.016	-140	0.9
U016	3	0.7	<0.05	<0.1	2.1	14.2	<0.001	-179	1.73
L113	3	0.05	<0.1	0.1	0.1	6.38	0.018	-123	-

(U, M, and L prefixes refer to monitoring wells screened in the upper, middle, and lower portions of the aquifer, respectively.)

plume in excess of the 1 nM normally associated with sulfate reducing conditions. It was therefore unclear whether natural attenuation was a considerable factor in the development and fate of the plume.

Mathematical modelling on BIOPLUME III showed that the plume has had adequate time to reach equilibrium. It also predicted that without biodegradation, the contaminants would reach the Mississippi River at concentrations approaching 10 ppm 4 to 5 miles downgradient of the contaminant source (Figure 2).

Twelve years of ground water data from a large number of monitoring wells demonstrate that the plume has never reached these proportions, and that at many locations concentrations are actually decreasing. In addition, the cDCE and 1,1-DCE concentrations cannot account for the loss of TCE and TCA in the ground water from the source area to the distal portions of the plume. The current configuration of the plume can only be explained by including some mechanism of *in situ* contaminant destruction.

Table 2. Contaminant concentrations in the ground water plume at TCAAP All concentrations are in μg/L. (See Figure 1 for well locations.)

Well	TCE	*c*DCE	*t*DCE	VC	TCA	1,1-DCA	1,1-DCE	BTEX	Ethene	Ethane
U020	588	4	<1	<1	446	12	57	<1	<1	<1
M020	108	1	<1	<1	3	3	1	<1	<1	<1
L020	61	<1	<1	<1	2	1	<1	<1	<1	<1
U002	40.4	<1	<1	<1	20	15	13	<1	<1	<1
M002	176	3	<1	<1	22	20	15	<1	<1	<1
L002	144	1	<1	<1	17	4	7	<1	<1	<1
U821	65	<1	<1	<1	8	2	3	<1	<1	<1
U872	121	1	<1	<1	11	5	7	<1	<1	<1
U016	ND	<1	<1	<1	<1	<1	<1	<1	<1	<1
L113	ND	<1	<1	<1	<1	<1	<1	<1	<1	<1

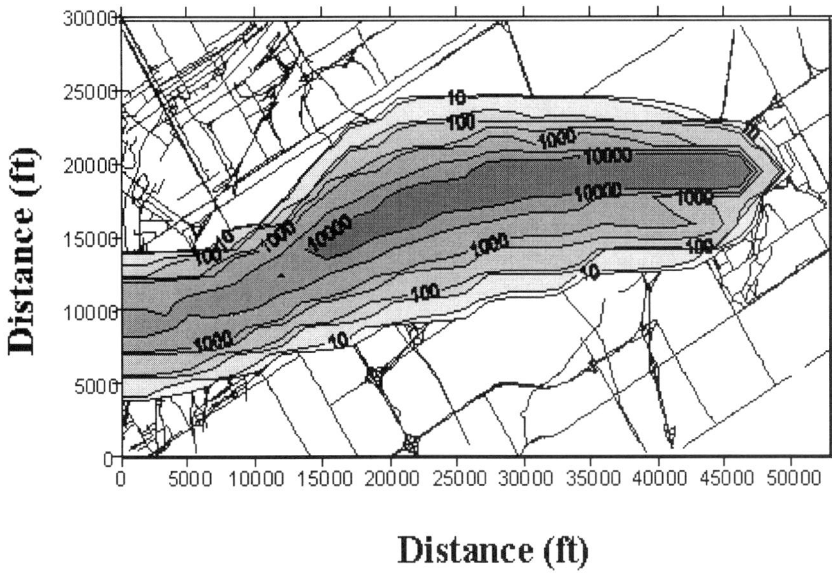

Figure 2. BIOPLUME III simulation of the TCAAP plume for 1990 assuming no biodegradation of TCE or TCA. The model assumes that the contaminant source begins in 1940. Concentration contours are in µg/L.

Figure 3 shows a regression of natural log transformed TCE concentration data as a function of distance from the on-site source. The slope of the regression can be viewed as the total attenuation rate divided by the ground water velocity (Buscheck and Alcantar, 1995). The ground water velocity was estimated at approximately 950 ft yr^{-1} in unrelated studies, yielding an estimate of the attenuation rate of -0.2 yr^{-1}. This rate is low in comparison to those found at other sites (Wilson et al., 1996), which can range from -0.1 to over -5 yr^{-1}.

BIOPLUME III accurately predicts the current configuration of the ground water contaminant plume when a biodegradation rate of –0.2 yr^{-1} is included (Figure 4) in the simulation. It predicts that TCE should currently be found at concentrations of less than 100 • g/L 21,000 feet downgradient of the source. This is supported by actual ground water data trends.

The mathematical modelling shows that biological transformation of TCE and TCA is required to explain the actual development and stabilization of the plume. The fate of the metabolic breakdown products, cDCE and 1,1-DCE is not understood; the fact that they are not accumulating with distance along the ground water flowpath is evidence that they are breaking down *in situ*. However, the lack

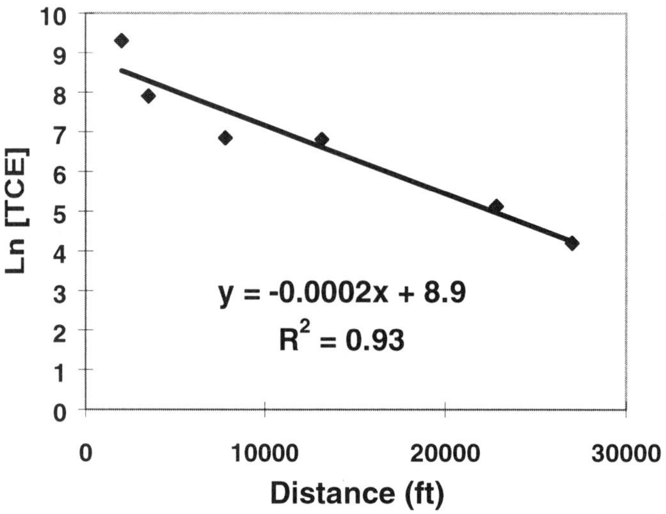

Figure 3. TCE concentration (µg/L) versus distance from the contaminant source. Multiplying the slope of the regression by the ground water velocity of approximately 1000 ft yr^{-1} gives an attenuation rate of –0.2 yr^{-1}

of VC and ethene suggest that cDCE and 1,1-DCE are not breaking down through reductive dechlorination. Recent studies (Bradley et al. 1998) show that bacteria in aquifers can degrade cDCE to carbon dioxide using manganese IV as the electron acceptor. This would explain why the reduced transformation products do not accumulate in the ground water at TCAAP.

Other studies (Jakobsen et al., 1998) show that ground water temperature may affect hydrogen concentrations, with cold ground water producing more hydrogen. The ground water at the TCAAP site is significantly colder (11 degrees C) than sites in warmer regions of the U.S. that serve as models for geochemistry versus hydrogen concentration. In addition, regional ground water is often manganese reducing but not iron reducing, an unusual phenomenon in ground water of the U.S. (Brandon et al., 1998). The association of specific concentrations of hydrogen with geochemistry and biodegradation processes may therefore depend on the location and geologic setting of the aquifer. Current microcosm studies are underway to determine the fate of cDCE and 1,1-DCE in the ground water at this site. Meanwhile, the estimate of the time required for ground water restoration at the site has decreased significantly based on the results of this study.

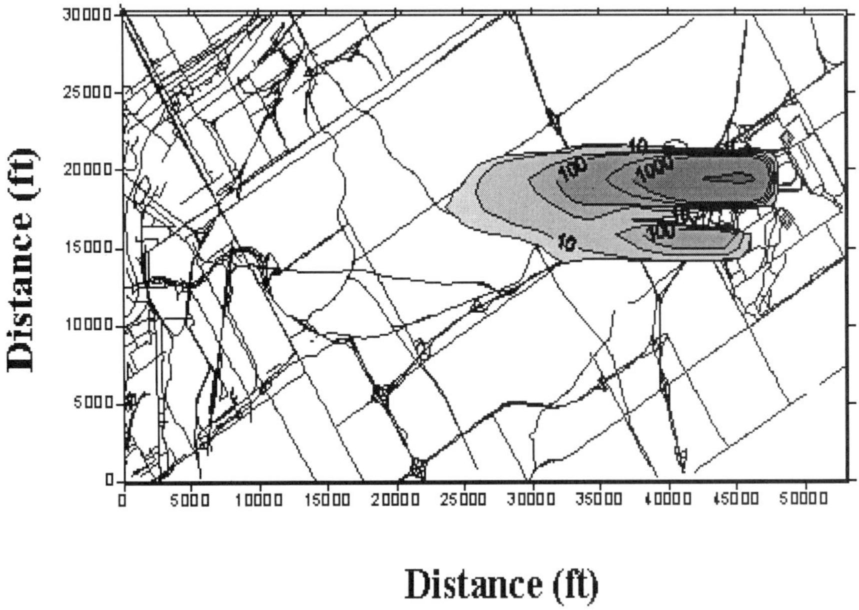

Figure 4. BIOPLUME III simulation for 1990 assuming a biodegradation rate of –0.2 per year. This simulation closely resembles the actual development of the plume. Concentration contours are in units of • g/L.

CONCLUSIONS

At this site, the hydrogen concentrations appear to be a better indicator of the potential for the biodegradation of TCE and TCA than the geochemical data alone. Although the redox status of the ground water is predominantly manganese or iron reducing, the hydrogen concentrations in excess of 1 nM indicated that the reductive dehalogenation of TCE and TCA is possible. Based on mathematical modelling, the degradation of TCE and TCA is an important factor in the development and fate of the ground water plume. The breakdown products, cDCE and 1,1-DCE, are not accumulating in the ground water and may be degrading, probably through a mechanism other than reductive dechlorination. This analysis may be important to natural attenuation studies of chlorinated solvents in glacial outwash aquifers.

DISCLAIMER

This paper has been reviewed in accordance with the U.S. Environmental Protection Agency's policies and has been approved for presentation and publication.

REFERENCES

Bradley, P. M., J. E. Landmeyer, and R. S. Dinicola. 1998. "Anaerobic oxidation of 1,2 (14)C-dichloroethene under Mn(IV)-reducing conditions." *Appl. Environ. Microbiol. 64*:1560-1562.

Brandon, B., M. Nalipinski, M. Ferrey, and P. Estuesta. 1998. "Measuring Mn+2 in ground water natural attenuation studies", In G. B. Wickramanyake and R. E. Hinchee (Eds.), *Natural Attenuation: Chlorinated and Recalcitrant Compounds*, pp. 193-198. Battelle Press, Columbus, OH.

Buscheck, T. E., and C. M. Alcantar. 1995. "Regression techniques and analytical solutions to demonstrate intrinsic bioremediation." In R. E. Hinchee, J. T. Wilson., and D. C. Downey, (Eds.), *Intrinsic Bioremediation*, pp. 109-116. Battelle Press, Columbus, OH.

Jakobsen, R., H.-J. Albrechtsen, M. Rasmussen, H. Bay, P. L. Berg, and T. H. Christensen. 1998. "H2 concentrations in a landfill leachate plume (Grindsted, Denmark): In situ energetics of terminal electron acceptor processes." *Environ. Sci. Technol. 32*:2142-2148.

Lovley, D. R., F. H. Chapelle, and J. C. Woodward. 1994. "Use of dissolved H2 concentrations to determine distribution of microbially catalyzed redox reactions in anoxic groundwater." *Environ. Sci. Technol. 28*:1205-1210.

U. S. Environmental Protection Agency. 1998. *Technical Protocol for Evaluating Natural Attenuation of Chlorinated Solvents in Ground Water*. EPA/600/R-98/128.

Vogel, T. M., and P. L. McCarty. 1985. "Biotransformation of tetrachloroethylene to trichloroethylene, dichloroethylene, vinyl chloride, and carbon dioxide under methanogenic conditions." *Appl. Environ. Microbiol. 49*:1080-1083.

Wilson, J. T., D. H. Kampbell, and J. W. Weaver. 1996. "Environmental chemistry and the kinetics of biotransformation of chlorinated organic compounds in ground water." *Proceedings of the Symposium on Natural Attenuation of Chlorinated Organics in Ground Water*, Dallas, TX, Sept. 11-13, pp 124-127.

AQUEOUS AND MINERALOGICAL INTRINSIC BIOREMEDIATION ASSESSMENT: INCENTIVE AND EXAMPLE

Lonnie Kennedy (Kennedy Geosciences, Norman, OK)
Jess Everett, Tina Conroy, Luis Garcia, Claire Steager (Rowan University, Glassboro, NJ)

ABSTRACT: Natural attenuation studies typically focus on evaluating electron acceptors or respiratory products dissolved in ground water. Little emphasis is placed on evaluating the solid aquifer matrix. However, most aquifer sediments contain a large oxidative capacity in the form of solid mineral iron (Fe^{3+}). Conversely, solid mineral species are formed as aqueous SO_4^{2-} and mineral Fe^{3+} are reduced. AMIBA (Aqueous and Mineralogical Intrinsic Bioremediation Assessment) uses aqueous and mineral assessment to: (1) more completely assess assimilative and expressed capacity; (2) use sediment concentrations of Fe^{2+} to mark the former extent of the plume; and (3) provide information for better predictive mass transport modeling. AMIBA was demonstrated at a fire training site on Westover Air Force Base where BTEX and chlorinated solvents have been found in ground water. BTEX is shown 97% remediated by intrinsic bioremediation alone. Though reductive dechlorination occurs, there is no buildup of vinyl chloride (VC). The three-dimensional distribution of bioavailable mineral Fe^{3+} was evaluated to show proper distribution and sufficient oxidative mass for microbial oxidation of VC by iron reduction.

INTRODUCTION

Natural attenuation is recognized as an effective remediation for hydrocarbon fuels and chlorinated solvents. This paper briefly describes the purpose, methods, results, and benefits of performing an Aqueous and Mineralogical Intrinsic Bioremediation Assessment (AMIBA) by including solid Fe and S mineral data as part of a thorough natural attenuation study.

Natural attenuation includes many abiotic processes, including advection, dispersion, volatilization, that collectively decrease contaminant concentrations. However, intrinsic bioremediation is often the chief destructive mechanism. Determining the efficacy of natural attenuation has been accomplished by performing natural attenuation studies that typically relying on aqueous data supplied from ground water analyses (AFCEE, 1995 and ASTM, 1995). With this approach it is possible to roughly identify operative redox zones and/or daughter products and thus confirm that remediation via microbial oxidation/reduction is likely occurring. However, at the conclusion of such a study, the historical efficiency of intrinsic bioremediation to ameliorate a site cannot be reasonably quantified. Thus, extended periods of monitoring are necessary to confirm plume stabilization and decay.

For an engineered treatment system, an efficiency study would be conducted to determine its effectiveness after some period of operation. Bioremediation is a de facto treatment system, operational since the first release. As such, intrinsic bioremediation should also be routinely examined for cleanup efficiency, which is only possible with an adequate site assessment similar to the one described here. AMIBA can produce a defendable value for the percent contaminant remediated by natural attenuation. It can also define the maximum historical extent of plume migration (plume footprint) to effectively demonstrate plume contraction.

With a traditional natural attenuation assessment, indirect data are often used to estimate dissolved phase degradation rate constants (Wiedemeier et al. 1996, Buscheck and Alcantar, 1995). Rate constants can be applied to a predictive model to forecast future plume distribution and decay. However, a great deal of uncertainty of remains concerning the rate of source decay and source loading, factors that strongly affect simulation outcomes. Indeed, knowing only an approximate dissolved phase degradation rate and given the conservative assumption of a constant source, one can at best only demonstrate a steady state plume in a simulation. An AMIBA study can aid in quantifying mass loading and source decay rates in support of predictive modeling.

It can be argued that the importance of natural attenuation is marginal, in light of the tendency to rely on Risk Based Corrective Action (RBCA) for site closure. However, success in RBCA does not infer that contaminants are absent from the site. In fact, RBCA relies exclusively on natural attenuation processes for complete site restoration if any level of contaminant is present. RBCA is strengthened when key input factors are supplied from a comprehensive natural attenuation assessment rather than relying on generalized data. Thus, the need for improved natural attenuation assessment is even more critical when RBCA is considered.

Finally, the level of effort and costs required for AMIBA may be greater than normally expended on a site assessment. Though not discussed here, methods have been developed to reduce sample frequency. AMIBA can demonstrate that intrinsic bioremediation: 1) has historically been effective in contaminant abatement; and 2) will continue to restore the site in a timely manner, so a more costly, engineered remediation system is not required. Further, knowing the intrinsic bioremediation efficiency (percent contaminant cleanup), source decay rate, and armed with footprint data showing plume retreat one could reduce or possibly eliminate the need for costly on-going monitoring if natural attenuation (or RBCA) is selected as the remedial option. On the other hand, if remediation is necessary, a good natural attenuation study will fully characterize the spatial distribution of the contaminants making remedial design significantly more effective.

With AMIBA, the role of natural attenuation in augmenting, or hindering, engineered treatment can be assessed. The concept that natural attenuation might inhibit engineered treatment is a fact not often recognized. For example, as shown in this study, intrinsic bioremediation processes can produce quite large amounts of reduced reactive mineral species that can strongly compete with bacteria as a sink for oxygen should sparging or the chemical addition of oxygen be considered as a treatment. Indeed, the chemical oxygen demand from minerals generated during natural attenuation can be many times greater than the targeted contaminant. Abiotic mineral demand cannot be determined by examining ground water data alone. Using AMIBA, intrinsic bioremediation and engineered treatment systems can be designed to work together to improve cleanup and reduce costs. At a minimum, natural attenuation should be relied upon to address any post treatment residual contaminants as part of life-cycle site management.

REOX PROCESSES

Complete mineralization of organic substrate is accomplished through a series of enzymatically-mitigated pathways, which ultimately can be described in terms of simple redox processes. In addition to oxygen, indigenous soil bacteria have the ability to oxidize organic contaminants using alternative electron acceptors including nitrate, sulfate, and iron. If a sufficiently large hydrocarbon spill occurs, a redox series may develop as electron acceptors are preferentially utilized. Common convention holds that the redox order is oxygen > nitrate > iron > sulfate > methanogenesis. There is ample evidence that, under aquifer conditions, mineral Fe^{3+} reduction occurs over a rather wide redox range extending through methanogenesis. The mineral form of the iron, surface area, chelating agents, contact time with the contaminant, contaminant type, dissolution kinetics and other factors may all be factors in the quantity and rate of Fe^{3+} utilization.

With respect to electron acceptor abundance, oxygen is typically limited by saturation constraints to less than 8 ppm in pristine shallow aquifers. Though highly soluble, nitrate is normally found in low concentrations, typically below the drinking water standard of 10 ppm $N\text{-}NO_3$. Conversely, sulfate is often found in concentrations exceeding 25 ppm. The (secondary) drinking water standard for sulfate is 250 ppm. Oxygen, nitrate, and sulfate are aqueous electron acceptors. Iron exists almost exclusively as a mineral that cannot be assessed using water analyses. However, iron is the fourth most abundant element in the Earth's crust and is very common in shallow geologic strata, about 0.1 to 100% in ore grade materials. Based on common aquifer characteristics, the ratio of oxidative capacity for nitrate, sulfate, and bioavailable iron relative to oxygen is about 1:4:10+ respectively. In other words, per unit volume aquifer material, sulfate or iron reduction could account for between four to greater than ten times as much organic contaminant oxidation than oxygen and nitrate. This suggests that sulfate and iron could easily dominate intrinsic bioremediation processes.

It must be recognized that soluble electron acceptors are resupplied to a contaminated area by ground water transport processes, whereas Fe^{3+} mass is fixed. The microbial utilization of oxygen, nitrate, and sulfate is largely facilitated by the movement of those electron acceptors to the contaminated source area. Alternatively, the contaminant must move to the Fe^{3+} before iron reduction can occur. Supplies of oxygen, nitrate, and sulfate are typically consumed near the source area (head of the plume), a condition that favors oxidation by Fe^{3+} reduction (and methanogenesis) of the long tail of a dissolved phase plume.

Additionally, if a plume contracts, a trail or "footprint" of reduced Fe^{2+} can be left behind marking the prior position of the plume. Plume contraction is indicated if the extant hydrocarbon plume is significantly smaller than the Fe^{2+} footprint.

Little definitive evidence remains when contaminants are oxidized using oxygen and nitrate. The respiratory products of oxygen and nitrate reduction are nitrogen gas, carbon dioxide, and water. Conversely, the respiratory product of sulfate reduction (HS^-) commonly reacts with mineral iron species to form new mineral precipitants summarized as iron monosulfide (FeS), elemental sulfur (S^o), and amorphous pyrite (FeS_2). In addition to the already mentioned iron sulfide minerals, the reduction of Fe^{3+} commonly precipitates minerals such as siderite ($FeCO_3$), magnetite (Fe_3O_4), and vivinite ($Fe_3(PO_4)_2 \bullet 8H_2O$). Thus, unlike oxygen and nitrate, the respiratory products of sulfate and iron reduction are preserved in mineral form. As the deposition of those minerals is continuous, a mineral assay will approximate cumulative iron and sulfide respiration.

Normally, the background ratio of dissolved oxygen/sulfate and nitrate/sulfate entering a contaminated zone is constant. Therefore, using AMIBA to determine total expressed sulfate facilitates a better estimation of expressed oxygen and nitrate. Next, given reasonable estimates of reduced oxygen, nitrate, sulfate, and iron it is possible to estimate the quantity of contaminant destroyed (expressed capacity) using a mass balance approach. Determining the efficiency of intrinsic bioremediation is then simply a matter of comparing expressed capacity with the existing hydrocarbon mass. Finally, knowing the expressed capacity, current hydrocarbon mass, and the approximate time of contamination permits the estimation of a first order rate constant for overall contaminant decay. This source decay term combined with the total hydrocarbon mass (expressed capacity + contaminant extant) can be applied to a computer simulation to improve simulation accuracy.

SITE DESCRIPTION

A demonstration study for AMIBA was conducted at Fire Training Area, FT-08, at Westover Air Reserve Base, in Chicopee Massachusetts. The site was used between 1964 and 1986 as a fire training exercise area and for underground waste oil storage. Fuels and solvents were released at the site. Several site assessments, a risk assessment, and a standard intrinsic bioremediation assessment were performed (AFCEE, 1997). Groundwater samples contained high levels of fuel hydrocarbon (TPH) and low levels of various chlorinated hydrocarbons.

The unconfined shallow aquifer consists of medium to coarse alluvial sediment with a water table about four feet below the surface. The ground water flow is curved but is generally towards the east or east-northeast and may change orientation seasonally. Figure 1 shows 1) the ground water flow direction, 2) source areas, 3) the location of monitoring wells and soil borings from this and prior studies, 4) the line of section A to A' used for all cross-section profiles exhibited here, and 5) dissolved phase BTEX concentrations.

METHODS

For this study, 18 soil borings were drilled, as shown on Figure 1, to depths of 36 feet. The sample collection, preservation, extraction, and analyses techniques used for iron and sulfide mineral analyses are described in Kennedy et al. (1998a). Briefly, a simple technique has been developed to rapidly collect and preserve anoxic sediments under N_2 gas. Biologically significant Fe^{3+} and reduced Fe^{2+} minerals are extracted using 0.5 M HCl. Extracted Fe total and Fe^{2+} are analyzed spectrophotometrically. Fe^{3+} is determined by subtracting Fe^{2+} from Fe total. The sulfide from mineral FeS is extracted using a 6N HCl solution. The same sediment sample is then subjected to a 1 N Cr^{2+} and 12 N HCl solutions which will extract undifferentiated mineral S^o and FeS_2. Extracted sulfides were analyzed spectrophotometrically. Sediment samples were collected about every two feet through highly contaminated intervals, but this distance was increased with depth and diminishing organic content based on field screening with a photoionizing detector. Two hundred soil samples were analyzed for iron and sulfide minerals. Pore water extracted from those samples was analyzed for sulfate using ion chromatography and for dissolved iron by spectrophotometry. One hundred and sixty sediment samples were evaluated for fuel and volatile organics using EPA Method 8260. Eleven of the existing monitoring wells were sampled for dissolved oxygen, nitrate, sulfate, iron, and organics as above but the remainder could not be found or were dry at the time of this study.

RESULTS

All data was analyzed geostatistically using GridStat (1998) software. This software can perform three-dimensional analyses of spatially distributed data based on variogram analyses. The resulting grid block can be viewed in three-dimensions, cross-sectionally, or it can be summarized through the z-plane and viewed as two-dimensional isopleths maps. Due to space limitations only cross-sections along transect A to A' are shown here (Figure 1).

A dissolved BTEX plume was found to extend from the source area to approximately RU-8. Soil BTEX was confined to the two source areas (Figure 1). Based on 3-D spatial analyses the mass of remaining fuel is 2,050 moles in soil and 34 moles dissolved in ground water (240 L total). Only 1.7% of BTEX is dissolved in ground water, demonstrating that fuel mass at this site cannot be estimated by aqueous sampling as has been found by previous studies (Kennedy and Hutchins, 1993).

Oxygen and nitrate show reductions in concentrations typical of fuel-contaminated sites. Based on monitoring well data, background concentrations respectively ranged from 9.5 and 3.0 mg/L, respectively, to less than 1 mg/L in the plume. Concentrations of dissolved Fe^{2+} from pore water analysis ranged from < 1 to ~ 150 mg/L in the impacted area (Figure 2). Fe^{2+} distribution indicates a downward ground water flow/mass transport gradient. Significant dissolved phase Fe^{2+} concentrations do not extend to the down gradient well RU-9.

Mineral Fe^{2+} was evaluated using a normalization and filtering process. First, the ratio of Fe^{2+}/Fe total was determined and spatially interpolated. The background ratio of Fe^{2+}/Fe total was no more than 0.5. The background fraction was then removed from the geostatistically generated model matrix so that only Fe^{2+} from the microbial reduction of contaminants remained, referred to here as biogenic Fe^{2+}. Fe^{3+} was determined as Fe total – Fe^{2+} and modeled without further processing. As shown in Figure 3, concentrations of mineral biogenic Fe^{2+} increased in the contaminant area while Fe^{3+} decreased (Figure 4). Over the entire site, 5,270 moles dissolved Fe^{2+} was found as compared with 1,600,000 moles in mineral biogenic form. Thus, only 0.3% Fe^{3+} expressed capacity can be found in ground water. A serious error would result if only aqueous data were used. Finally, dissolved organic concentrations stop just beyond RU-8, yet very high concentrations of mineral biogenic Fe^{2+} extend well beyond RU-9. This iron footprint defines past plume extent and verifies its contraction over time.

The average up gradient concentration for sulfate is 25.5 mg/L, though higher concentrations were found in isolated pockets. Sulfate was essentially depleted via sulfate reduction in the center of the plume (Figure 5). Little sulfide from mineral FeS was found however, there were significant concentrations of sulfide from FeS_2 + S^o (Figure 6). Just 7,000 moles aqueous expressed sulfate were identified. However, mineral analyses identified 19,100 moles expressed sulfate. The traditional aqueous assessment identified only 37%.

Normally, expressed oxygen and nitrate would be determined through aqueous analyses with about the same degree of underestimation error as was found for aqueous sulfate above. However, expressed oxygen and nitrate can be more accurately determined by indexing to reduced mineral sulfide. This method recognizes microbes only consume sulfate after oxygen and nitrate are depleted. Further, the ratio of background oxygen/sulfate and nitrate/sulfate must be determined (1.2:1 and 0.8:1 respectively at this site) and the mass of expressed sulfate from mineral sulfide must be known accurately. Thus, for every mole sulfate consumed 1.2 moles of oxygen and 0.8 moles nitrate were expressed or 22,700 m and 15,300 m respectively. For this technique it is not necessary to directly measure oxygen or nitrate spatially throughout the impacted aquifer region. One must merely confirm ΔO_2 and ΔNO^{3-} across the site and accurately determine expressed SO_4^{2-} from mineral analyses.

With the reasonable estimates for expressed oxygen, nitrate, sulfate, and iron found via AMIBA, it is possible to calculate the amount of BTEX consumed using a standard chemical mass balance approach as in Wiedemeier et al. (1996) or Kennedy et al. (1998). Based on this analyses it can be demonstrated that at least 52,900 moles (6,100 L) BTEX has been biodegraded. Knowing the existing BTEX mass, 2,100 moles (242 L), it is possible to state that at least 97% of the BTEX has been remediated by intrinsic bioremediation processes alone; an impressive treatment achievement for any technology.

With AMIBA, predictive mass transport computer modeling can be more accurately performed because an estimate of total fuel release (biodegraded + existing) is now known. Further, knowing the total released fuel, current fuel, and an estimate of the time of release it is possible to calculate biodegradation rate kinetics for the source to further support computer modeling.

Several previous investigations found chlorinated compounds at this site, though concentrations were decreasing (AFCEE 1997). This study found only dissolved phase trichloroethylene and cis-1,2-dichloroethylene, at maximum concentrations of 6640 ug/L and 2290 ug/L respectively at well CS-3. There is significant evidence of reductive dechlorination without the buildup of vinyl chloride (VC). Iron reducing bacteria can rapidly oxidize VC (Bradley and Chapelle, 1996). As illustrated in Figure 4, this site easily contains a large surplus of Fe^{3+} sufficient to meet the assimilative demands of much larger quantities of chlorinated solvents than have been reported at this site. Further, using AMIBA the spatial distribution of residual Fe^{3+} can be definitively mapped to show that it is in the proper redox zone and flow path to facilitate oxidation of VC.

REFERENCES

AFCEE, 1995. *Technical Protocol for Implementing Intrinsic Bioremediation with Long-Term Monitoring for Natural Attenuation of Fuel Contamination Dissolved in Groundwater.* AFCEE Technology Transfer Division, Brooks, AFB, San Antonio, TX.

AFCEE, 1997. *Treatability study in support of intrinsic remediation option at the current fire training area*, Westover Air Reserve Base, Chicopee, Massachusetts, Air Force Center for Environmental Excellence, Technology Transfer Division, Brooks Air Force Base, San Antonio, TX

ASTM, 1995. *Standard guide for risk-based corrective action (RBCA) applied at petroleum release sites, American Society for Testing and Materials*, E-1739-95.

Bradley, Paul M. and Francis H. Chapelle, 1996. "Anaerobic mineralization of vinyl chloride in iron (III) reducing aquifer sediments", *Symposium on Natural Attenuation of Chlorinated Organics in Ground Water*, EPA/540/R-96/509.

GridStat, 1998. *GridStat Users Manuel*, Applied Computer Engineering, Inc., Houston TX.

Kennedy, L., J. Everett, K. Ware*, and V. Green*, R. Parsons (1998a) "Iron and sulfur mineral analysis methods for natural attenuation assessments", *Bioremediation Journal*, 2(3&4):259-276.

Kennedy, L., J. Everett, T. Dewers, W. Pickins, D. Edwards (1998b) "Application of mineral iron and sulfide analysis to evaluate natural attenuation at a fuel contaminated site", *Journal of Environmental Engineering, ASCE*, 125(1).

Kennedy, Lonnie G. and Stephen R. Hutchins, 1993. "Applied Geologic, Microbiological, and Engineering Constraints of In-Situ BTEX Bioremediation", *Remediation*, Winter pp 83 -107.

Wiedemeier, T.H., M.A. Swanson, J.T. Wilson, D.H. Kampbell, R.N. Miller, and J.E. Hansen, 1996. "Approximation of biodegredation rate constants for monoaromatic hydrocarbons (BTEX) in groundwater", *Ground Water Monitoring and Remediation.*

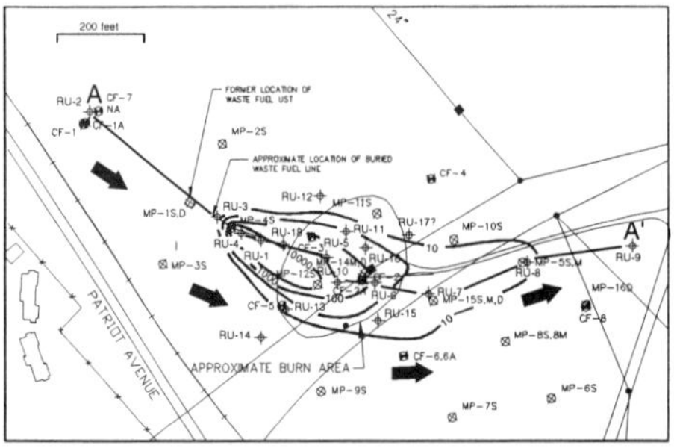

Figure 1. Site location map with BTEX (ug/L).

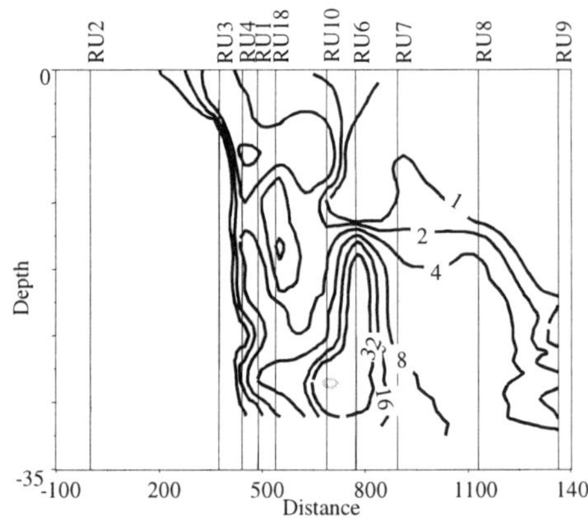

Figure 2. Pore water Fe (mg/L)

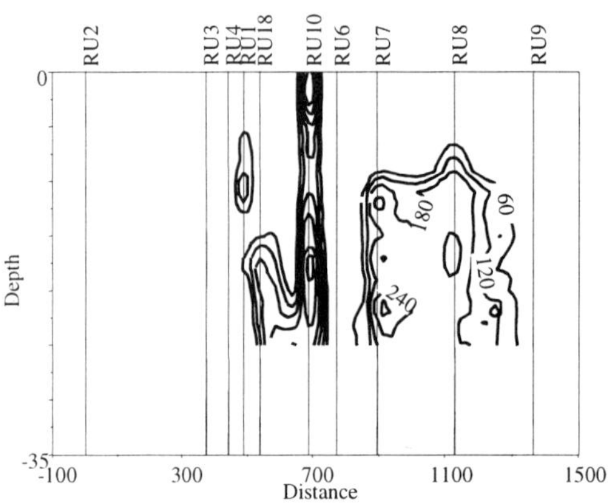

Figure 3. Net mineral Fe^{2+} (mg/kg)

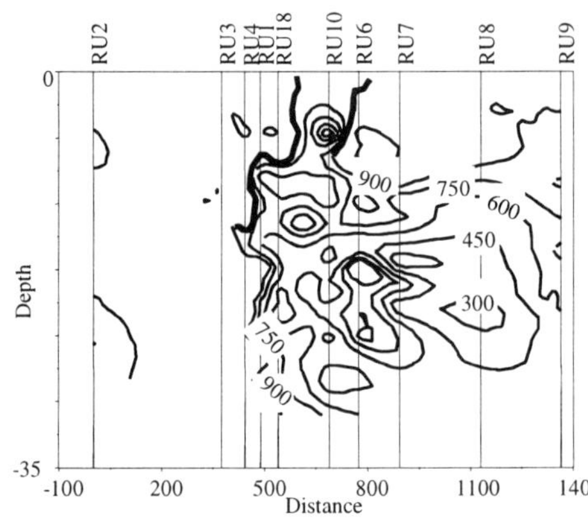

Figure 4. Mineral Fe^{3+} (mg/Kg)

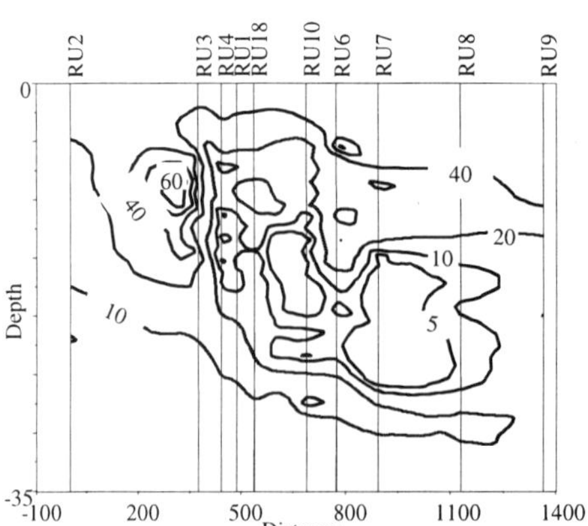

Figure 5. Pore water sulfate (mg/L)

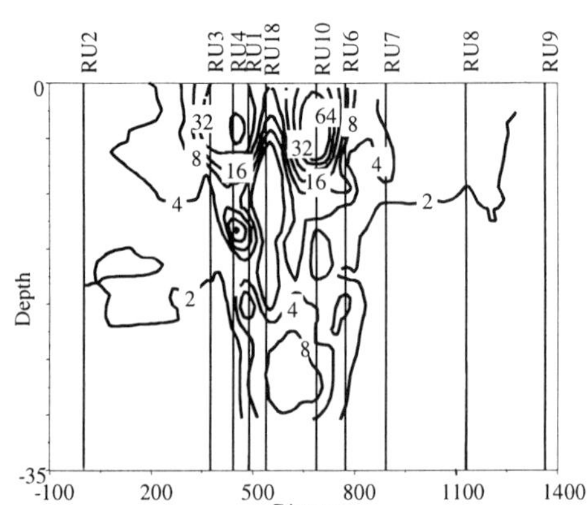

Figure 6. Chromium extractable mineral sulfide (mg/Kg)

CASE STUDY: ENVIRONMENTAL COMPIANCE BY NATURAL ATTENUATION OF MIXED CHLORINATED AND AROMATIC HYDROCARBONS

Franz K. Hiebert, *(RMT, Inc., Austin, Texas)*
John A. Miller (Schlumberger, Sugar Land, Texas)
Scott Madill (Radian International, LLC, Midland, MI)

ABSTRACT: Multiple lines of evidence indicate that in localized regions of an aquifer associated with a pre-existing acid plant and oil/water separator, groundwater conditions favor biodegradation of dissolved-phase chlorinated solvents and hydrocarbons.

The investigated site is an oilfield service facility in Enid, Oklahoma, which provided specialty products and services for cementing and stimulating oil and gas wells. Groundwater is impacted by low concentrations of dissolved-phase fuel hydrocarbons and chlorinated solvents. Following source removal, three years of quarterly sampling indicated that concentrations of dissolved-phase hydrocarbons and chlorinated hydrocarbons were declining in both mass and horizontal extent.

Measurements from 8 groundwater wells were collected and include: temperature, pH, reduction-oxidation (redox) potential, dissolved oxygen, carbon dioxide, nitrate, sulfate, methane, ethane, ethene, total organic carbon, chloride, conductivity, and aerobic and anaerobic microbiology. Most measurements were made in-situ to preserve redox conditions. Samples sensitive to oxygen were collected under anaerobic conditions. Microorganisms were identified to species and strain level.

Chemical and microbiological data from across the site indicate a nested pattern of subsurface geochemical zones conducive to biodegradation of aromatic fuel hydrocarbons and reductive dehalogenation of trichloroethene (TCE).

A quarterly monitoring program provides data to verify that the concentration of dissolved-phase constituents in groundwater continues to decline, and that natural biodegradation of dissolved-phase constituents will continue to act as a remediation mechanism.

In 1999, monitored remediation by natural attenuation was accepted as the appropriate corrective action at this site by the Oklahoma Department of Environmental Quality.

INTRODUCTION

Remediation by natural attenuation (RNA) is a technology for reduction of low-concentration constituents in the environment. The term "natural attenuation" refers to naturally-occurring processes in soil and groundwater environments that act without engineering intervention to reduce the mass, toxicity, mobility, volume and concentration of contaminants in those media. The focus of this investigation was on the evaluation of biodegradation of mixed hydrocarbons and chlorinated hydrocarbons in groundwater at an oilfield service facility in Enid, OK. The facility provided specialty products and services for cementing and stimulating oil and gas wells in the region.

Since August 1996, environmental investigation and remediation activities have been performed at the site. A perched aquifer was identified at approximately 17-27 feet below grade, and a deeper aquifer at depths of approximately 36-46 feet below grade. Groundwater flows to the southeast. Soil containing fuel hydrocarbons and chlorinated solvents was removed to the depth of 15 feet in the vicinity of an oil/water separator. Sixteen monitoring wells were installed for sampling and evaluation of groundwater. Monitor wells 1S and 2S were installed near the former sources of gasoline and chlorinated solvents (Figure 1a). Monitor well 4S is located up-gradient of the source area and data from this well is used to represent background conditions.

The facility is underlain by interbedded fine to medium grained, silty sands, firm sandy clays and silty clays and fine to medium grained gravels. Shale occurs beneath these deposits extending to a depth of 1,000 feet, and acts as an aquitard.

Residual concentrations of benzene, ethylbenzene, xylene, toluene, and chlorinated hydrocarbons related to perchloroethylene (PCE) were detected in groundwater near the removed soil. The extent of the impact to groundwater was delineated and groundwater quality monitored for three years. The concentration of dissolved-phase benzene, toluene, ethylbenzene, xylenes, (BTEX constituents) and dissolved-phase chlorinated hydrocarbons remained stable or declined since sampling events were initiated in February 1997.

The generally steady and declining concentrations of VOCs and a lack of significant migration of dissolved-phase constituents over the past few years suggested that mechanisms of natural attenuation are active at this site (Weaver et al., 1995; McCarty and Semprini, 1994), and indicated that an evaluation of natural attenuation was warranted.

Methods. Groundwater from selected existing wells was collected for geochemical and microbiological measurements. Analytical methods and parameters which were analyzed are presented in Table 1.

Most measurements were made in-situ to preserve redox conditions. Samples sensitive to oxygen were collected under anaerobic conditions. Microorganism species type was determined by carbon utilization tests and gas chromatographic analysis of cell fatty acids (Hemming, 1988). Species density in each sampled well was determined by optical density (Milke et al., 1993).

Results. A diverse indigenous community of mixed aerobic and anaerobic bacteria was identified in water sampled from each of the eight shallow groundwater wells (Table 2).

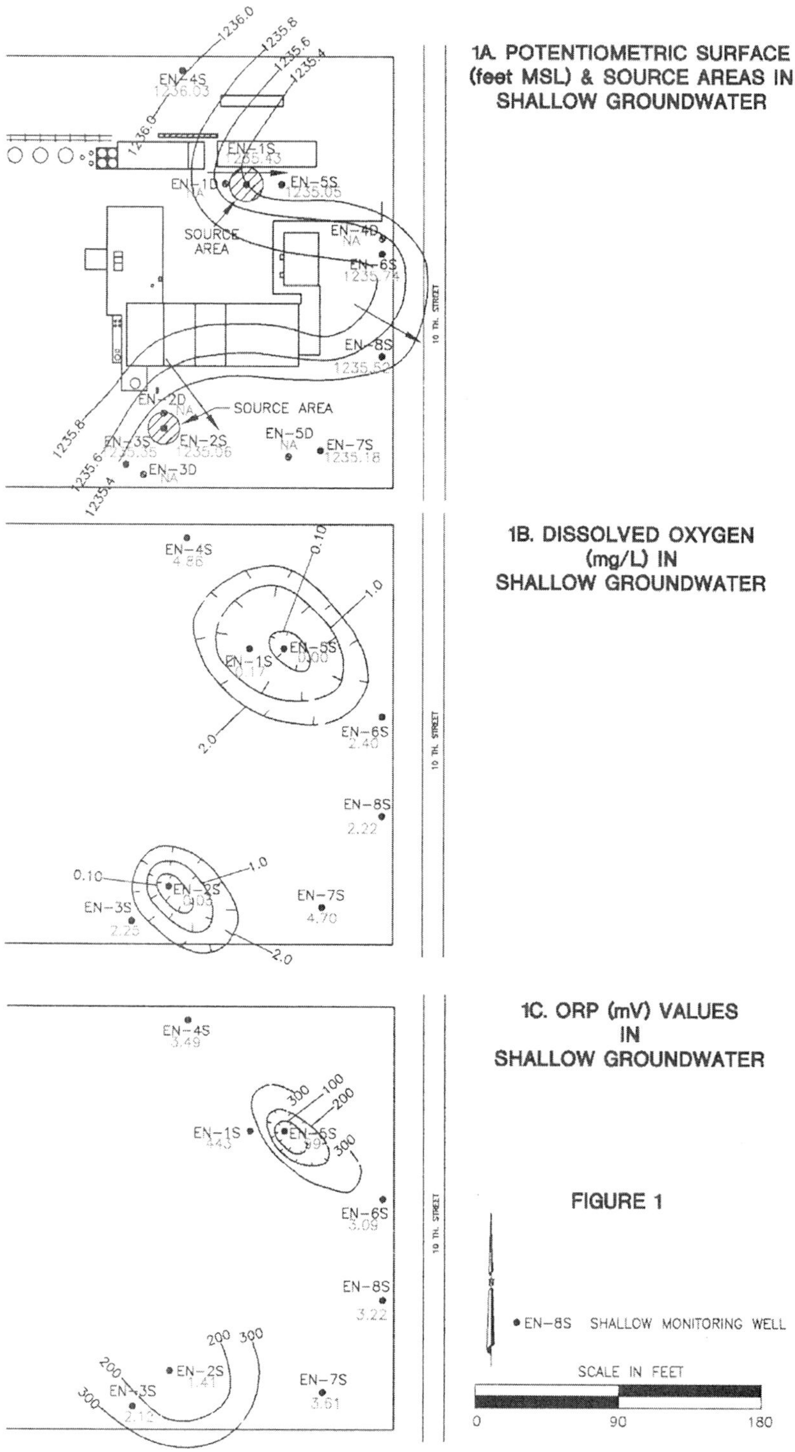
1A. POTENTIOMETRIC SURFACE (feet MSL) & SOURCE AREAS IN SHALLOW GROUNDWATER
1B. DISSOLVED OXYGEN (mg/L) IN SHALLOW GROUNDWATER
1C. ORP (mV) VALUES IN SHALLOW GROUNDWATER
SOURCE AREA
10 TH. STREET
EN-8S SHALLOW MONITORING WELL
SCALE IN FEET
0
90
180

FIGURE 1

Table 1. Evaluation of Natural Attenuation/Intrinsic Measurements Enid, Oklahoma

Parameter	Method
D.O., pH, ORP, Temp., S.C.	in-situ probe
Nitrate	USEPA 353.1
Sulfate	USEPA 375.4
Ferrous Iron	USEPA 315B
Chloride	USEPA 325.3
Total Organic Carbon	USEPA 415.1
Carbon Dioxide	AM15.01G (Microseeps, Inc.)
Methane	AM18G (Microseeps, Inc.)
Ethane	AM18G (Microseeps, Inc.)
Ethene	AM18G (Microseeps, Inc.)
Plate counts: bacteria	Milke *et al.*, 1993
Strain Identification	Hemming, 1988
Biodegradation	Milke *et al.*, 1993

Note: AM methods are modifications of Robert S. Kerr Laboratory SOP #114 (1991)

Table 2. Population of Bacteria in Shallow Groundwater By Well After 48 Hours of Incubation December, 1998

Total Heterotrophic Plate Count Results: Aerobic & Anaerobic

Direct Count: Colony Forming Units (CFU/ml) on 1/4 TSA

	Aerobic		Anerobic	
Sample	48 Hours	Types	48 hrs	Types
EN-1S	3.00×10^1	3	<10	0
EN-2S	1.70×10^2	6	confluent	1
EN-3S	4.80×10^2	4	1.60×10^2	3
EN-4S	3.98×10^4	8	1.08×10^3	7
EN-5S	1.00×10^2	2	<10	0
EN-6S	5.12×10^3	7	9.00×10^3	8
EN-7S	4.46×10^4	5	2.00×10^2	2
EN-8S	4.09×10^4	8	1.00×10^3	1

Confluent: Bacterial growth spreading across the plate, making accurate counts impossible.

Chemical parameters of sampled groundwater generally showed evidence of microbial activity in the subsurface (Table 3). pH levels in shallow groundwater ranged from 3.47 to 7.13 standard units. The lowest value occurred near the former acid plant source area and all others were circum-neutral. Concentrations of dissolved oxygen (D.O.) in groundwater ranged from 0.00 to

4.86 mg/L. Localized areas of anoxic, shallow groundwater were identified in and around both source areas (Figure 1b). Oxidation reduction potential (ORP) values in shallow groundwater ranged between -101 and 243 millivolts (mV) (Figure 1c).

Nitrate in shallow groundwater ranged from <0.05 to 12 mg/L. Compared to measured concentrations in the background well, 4S, nitrate and sulfate were depleted in the source area wells, 1S and 2S (Table 3).

Carbon dioxide, measured as a gas in groundwater, was detected in all of the sampled wells, and ranged from 35 to 554 mg/L. Elevated levels of CO_2 were measured in both source area wells, and in downgradient well EN-5S (Figure 2a).

Methane, as a gas within groundwater, was detected in all wells sampled, and ranged from 0.539 to 3100 ug/L in shallow screened wells. Methane concentrations in shallow groundwater were similar in spatial distribution to the CO_2 at the site, with elevated levels, compared to background, occurring in both source areas (Figure 2b).

Ethene ranged from <5.0 to 14016 ng/L (Figure 2c). Ethane ranged from <5.0 to 248 ng/L (Table 3). Both of these light hydrocarbons were distributed across the site in a pattern similar to carbon dioxide and methane.

Discussion. The preponderance of the data collected at this site support the interpretation that hydrocarbons and chlorinated hydrocarbons are naturally biodegrading.

Environmental conditions in the groundwater at the facility are favorable for the process of reductive dechlorination. Redox conditions, as measured by oxidation-reduction potential (ORP), are mostly positive across the site with areas of localized negative values in both source areas in shallow groundwater, indicating localized reducing conditions. Redox reactions in organic-rich groundwater are usually the result of metabolic activity of microorganisms, and the redox potential serves as a guide to interpreting the level and type of microbial activity in the sampled water (Wiedemeier et al., 1995).

Methane levels in both source areas in shallow groundwater represent an indicator of geochemical conditions required for microbial degradation of chlorinated hydrocarbons. Elevated methane concentrations are direct evidence of reducing geochemical conditions required for the process of reductive dehalogenation.

Carbon dioxide is an end product of biodegradation of hydrocarbons and certain chlorinated hydrocarbons. The spatial correlation between depressed concentration of D.O. and elevated CO_2 indicates that anaerobic metabolism is possible in the subsurface in this area.

Ethene is a byproduct of the biodegradation of PCE and TCE. Elevated levels of ethene were measured in both source areas in shallow groundwater, and solvent plumes typified by reducing conditions, and low concentrations of D.O., nitrate, sulfate and elevated concentrations of methane and FeII (not measured in this study) (Wiedemeier et al. 1999, p. 270).

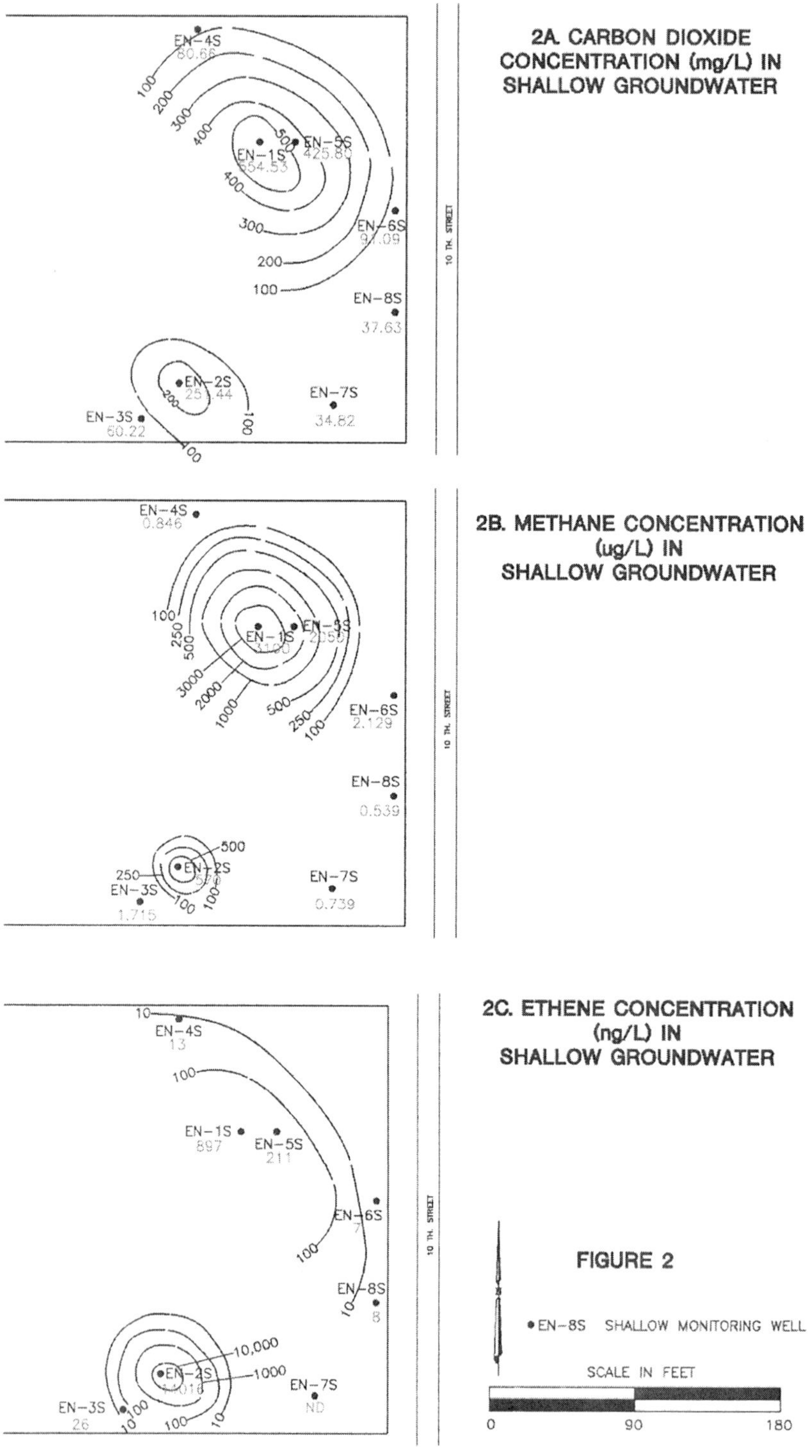

2A. CARBON DIOXIDE CONCENTRATION (mg/L) IN SHALLOW GROUNDWATER
EN-4S
80.66
EN-1S
EN-5S
EN-6S
EN-8S
37.63
EN-2S
EN-7S
34.82
EN-3S
60.22
10 TH. STREET
2B. METHANE CONCENTRATION (ug/L) IN SHALLOW GROUNDWATER
EN-4S
0.846
EN-1S
EN-5S
EN-6S
2.129
EN-8S
0.539
EN-2S
EN-7S
0.739
EN-3S
1.715
10 TH. STREET
2C. ETHENE CONCENTRATION (ng/L) IN SHALLOW GROUNDWATER
EN-4S
13
EN-1S
897
EN-5S
211
EN-6S
7
EN-8S
8
EN-2S
EN-7S
ND
EN-3S
26
10 TH. STREET
FIGURE 2
EN-8S SHALLOW MONITORING WELL
SCALE IN FEET
0
90
180

Table 3. Chemical Characterization of Groundwater December, 1998

Monitor Well	T (°C)	pH (s.u.)[1]	ORP[2] (mV)[3]	D.O.[4] (mg/L)	S.C. (mS/cm)[5]	Nitrate (mg/L)
EN-1S	19.58	3.47	243	0.17	14.68	<0.05
EN-2S	20.48	6.35	-59	0.03	7.84	<0.05
EN-3S	19.52	6.94	12	2.25	1.81	12
EN-4S	18.46	6.91	149	4.86	1.19	3
EN-5S	21.48	6.24	-101	0.00	8.17	<0.05
EN-6S	19.56	6.84	109	2.40	1.08	8.5
EN-7S	18.50	7.13	161	4.70	0.76	7.3
EN-8S	19.18	7.12	122	2.22	0.78	3.7

Monitor Well	Sulfate (mg/L)	T.O.C.[6] (mg/L)	Chloride (mg/L)	CO_2 (mg/L)	Ethane (ng/L)[7]	Ethene (ng/L)	Methane (ug/L)[8]
EN-1S	19	460	4200	554.53	6	897	3100
EN-2S	41	83	2300	251.44	248	14016	570
EN-3S	80	2.3	270	60.22	19	26	1.715
EN-4S	52	34	86	80.66	<5	13	0.846
EN-5S	18	24	2500	425.80	26	211	2050
EN-6S	49	<2.0	38	91.09	<5	7	2.129
EN-7S	38	2.3	10	34.82	10	<5	0.739
EN-8S	37	<2.0	24	37.63	<5	8	0.539

Notes:

1. Standard Units
2. Oxidation Reduction Potential
3. millivolts, silver/silver chloride reference potential
4. Dissolved Oxygen
5. Specific Conductance in milliSiemans per centimeter
6. Total Organic Carbon
7. Nanograms per Liter
8. Micrograms per Liter

suggest that the process of reductive dechlorination of chlorinated hydrocarbons is actively occurring in these areas (Wiedemeier et. al 1995). Ethane, a byproduct of the reductive dehalogenation of TCA was also measured at elevated concentrations in the same locations as ethene in shallow groundwater. This data suggests that biodegradation of chlorinated hydrocarbons is an active process in these areas.

The distribution of microorganisms in the groundwater at the facility indicates that a subsurface microbial community has adapted to the groundwater environment affected by the introduction of hydrocarbon and chlorinated solvents.

The spatial patterns of geochemical data collected at this site generally indicate that the same areas near well 1S and 2S behave as Type 1 chlorinated

Conclusions. Geochemical measurements of groundwater indicate that conditions currently favor biodegradation of PCE and residual fuel hydrocarbons. Spatial patterns of dissolved-phase constituents across the site also suggest a subsurface environment conducive to the long-term biodegradation of hydrocarbon and chlorinated constituents.

Note: In 1999, monitored remediation by natural attenuation was accepted as appropriate corrective action at this site by the Oklahoma Department of Environmental Quality.

REFERENCES

Hemming, B. C., 1988, Monitoring and identification of bacteria from agricultural environments by gas-chromatography fatty acid methyl ester (GC-FAME) analysis, in *Bioinstrumentation*, C. T. Lange, ed., pp. 57-70, Mathematics and Science Education Center, University of Missouri-St. Louis.

Hemming, B.C., and Milke, J. K., 1996, Working with the microbiological laboratory: insight into biofeasibility test procedures and bacteriological methods, Remediation Management, Vol. 2, No. 5, 1996, pp. 38-41.

McCarty, P.L., Semprini, L., 1994, Groundwater Treatment for Chlorinated Solvents, Handbook of Bioremediation, Norris et al., pp. 5-1 to 5-30.

Milke, J.K., Hemming, B. C., McCay-Buis, T., Huber, D. M., Lytle, F., 1993, Characterization of hydrocarbon degrading bacteria using a multi-well plate format, Proc. Am. Chem. Soc., Atlanta, GA., Sept. 28-30, 1993.

Weaver, J.W., Wilson, J.T., Kambell, D.H., 1995, Project Summary - Natural Attenuation of Trichloroethylene at the St. Joseph MI, Superfund Site, EPA/600/SV-95/001.

Wiedemeier, T. M., Swanson, M.A., MouFoux, D.E., Gordon, E.K., Wilson, J. T., Wilson, B.H., Kambell, D. H., Hansen, J.E., Haas, P., Chapelle, F.H., 1995, "Technical Protocol For Evaluating Natural Attenuation of Chlorinated Solvents in Groundwater." Air Force Center for Environmental Excellence, San Antonio, Texas.

Wiedemeier, T. H., Swanson, M.A., Wilson, J. T., Kambell, D. H., Miller, R. N., Hansen, J. E., 1995, Patterns of intrinsic bioremediation at two US Air Force Bases, in Intrinsic Bioremediation, R. E. Hinchee, J. T. Wilson, and D. C. Downey eds., Battelle Press, Columbus. pp. 31-52.

Wiedemeier, T. H., et al., 1999, Natural Attenuation of Fuels and Chlorinated Solvents in the Subsurface, John Wiley and Sons, Inc., New York, pp. 615.

NATURAL ATTENUATION OF CHLORINATED SOLVENTS – A CASE STUDY

Jennifer H. Lee and John A. Jackson (Parsons Engineering Science, Inc., Pasadena, California)
E. Kinzie Gordon and Todd H. Wiedemeier (Parsons Engineering Science, Inc., Denver, Colorado)
Michael J. Pound (Southwest Division, Naval Facilities Engineering Command, San Diego, California)

ABSTRACT: The occurrence of natural attenuation, primarily via biodegradation, was demonstrated for groundwater impacted by high concentrations of chlorinated aliphatic hydrocarbons (CAHs) and benzene, toluene, ethylbenzene, and xylene(s) (BTEX) beneath a golf course at Naval Air Station (NAS) North Island in San Diego County, California. The results of four rounds of quarterly groundwater sampling, along with available historical data, indicate that the areal extent of the volatile organic compound (VOC) plume, and the concentrations of parent CAHs within it, have decreased significantly since 1996. The distributions of geochemical indicators strongly suggest that anaerobic degradation of BTEX has created strongly reducing groundwater conditions conducive to the reductive dechlorination of CAHs. The presence of ethene and ethane, the final daughter products of reductive dechlorination, within the plume provided conclusive evidence that CAHs are degrading to innocuous end products. The groundwater quality and geochemical data also indicated that rapid destruction of vinyl chloride via aerobic biodegradation may be occurring at the downgradient edge of the plume. The results of this study contribute to the growing body of evidence that indicates that monitored natural attenuation (MNA) can be a valuable component of remediation for CAH-contaminated sites where receptor exposure points are outside of plume boundaries.

INTRODUCTION

Objective. A study was conducted for the Naval Facilities Engineering Service Center (NFESC) to assess the potential for MNA to reduce concentrations of CAHs and BTEX detected in groundwater beneath the Golf Course Disposal Area (Site 5, Unit 2) at NAS North Island in San Diego County, California. The objective of this evaluation was to determine if natural attenuation of dissolved chlorinated solvents and fuel hydrocarbons is occurring at the site, and if so, to investigate if the attenuation processes are sufficient to minimize the expansion of the contaminant plume and to ensure that federal and state groundwater protection standards are met at a downgradient point of compliance (POC).

Site Description. Site 5 is a former cut-and-cover sanitary landfill that received wastes, including hazardous materials, from the mid-1940s through 1965. The land then was operated as a waste transfer station until 1983, and was redeveloped as a golf course during 1983 and 1984. The source of groundwater contamination at the site is believed to be two former rectangular hazardous waste disposal pits located south of the former main landfill. High concentrations of VOCs, including tetrachloroethene (PCE), trichloroethene (TCE), dichloroethene (DCE), vinyl chloride (VC), and BTEX, were detected in the shallow aquifer.

Shallow soils consist of 5 to 10 feet of hydraulic fill material, composed of fine silty sand. In the southern portion of the site, the hydraulic fill is underlain by beach sands to a depth of 15 feet below ground surface (bgs). Beneath the hydraulic fill material and beach sand is a 3- to 5-foot-thick zone of low-permeability silt and clay that separates the overlying sediments from the underlying Bay Point Formation. The depth to groundwater ranges from approximately 6 to 12 feet bgs. Groundwater flow is primarily to the southwest, with a less significant northwestern flow component beneath the southern portion of the site (Figure 1). The secondary flow component is likely reflective of the local mounding effect caused by heavy irrigation of the golf course putting green located south of the site.

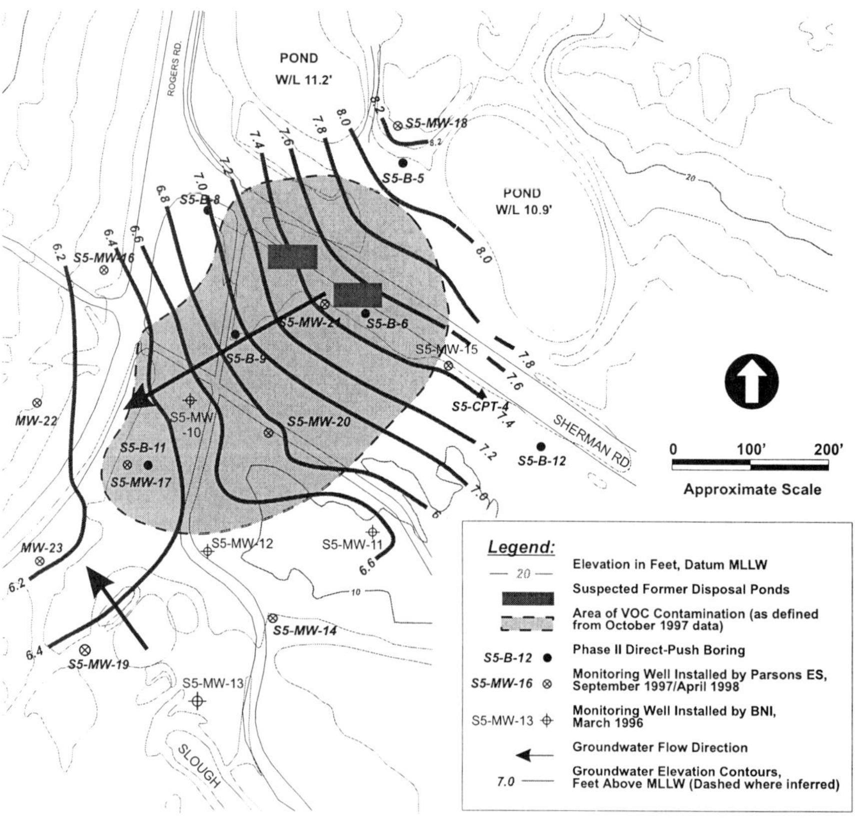

FIGURE 1. Site plan showing groundwater flow direction.

A stormwater drainage slough receives drainage via an 81-inch, north/south storm sewer located to the west of the site, and conveys it southward to the Pacific Ocean. This slough was identified as a POC.

DATA COLLECTION

Historical soil and groundwater data were available from a previous site investigation in 1996. To meet the requirements of a natural attenuation evaluation, additional data were required to determine aquifer properties, groundwater geochemistry, and the spatial and temporal behavior of the VOC plume.

Field activities conducted for this investigation included the installation of ten additional groundwater monitoring wells to supplement four existing wells at the site, and aquifer slug testing at two well locations to estimate the hydraulic conductivity of the shallow aquifer. Sampling and analysis activities were conducted in four quarterly events between October 1997 and July 1998, and included the collection of groundwater quality, geochemical, and elevation data necessary to assess the patterns of natural attenuation at the site. Groundwater geochemical parameters, including electron acceptors (dissolved oxygen [DO], nitrate, and sulfate), metabolic byproducts (ferrous iron [Fe^{+2}], carbon dioxide, chloride, methane, ethane, and ethene), and other indicators of biodegradation (dissolved hydrogen [H_2], oxidation/reduction potential [ORP], and alkalinity) were analyzed using field and fixed-base laboratory methods. The analyses were chosen and performed in accordance with the United States Environmental Protection Agency (USEPA) *Technical Protocol for Evaluating Natural Attenuation of Chlorinated Solvents in Ground Water* (Wiedemeier *et al.*, 1998).

RESULTS AND DISCUSSION

Multiple distinct but converging lines of evidence have been used in recent years to evaluate the efficacy of MNA (National Research Council, 1993; USEPA, 1997; Wiedemeier and Chapelle, 1998). Two of the most common lines of evidence used to document the occurrence of natural attenuation are: 1) historical trends in contaminant data showing plume stabilization and/or loss of contaminant mass over time, and 2) chemical and geochemical evidence of biodegradation. These lines of evidence were applied to the site as described below.

Documented Loss of Contaminant Mass or Plume Stabilization. In order to provide a rough basis of comparison for actual plume behavior, contaminant transport velocities for a no-biodegradation scenario were calculated using a hydraulic conductivity value of 37 feet per year (ft/yr), derived from aquifer slug tests, and assumed values of effective porosity, fraction soil organic carbon, bulk density, and total porosity representative of site conditions. Average transport velocities for PCE, TCE, *cis*-1,2-DCE, and VC were calculated to range from 54 ft/yr to 194 ft/yr. Accordingly, a plume consisting of these compounds would have been expected to migrate at least 63 feet (54 ft/yr x 1.17 yr) from August 1996 to October 1997, and at least 104 feet (54 ft/yr x 1.92 yr) from August 1996 to July 1998. Figure 2, which presents the extents of the total VOC plume as

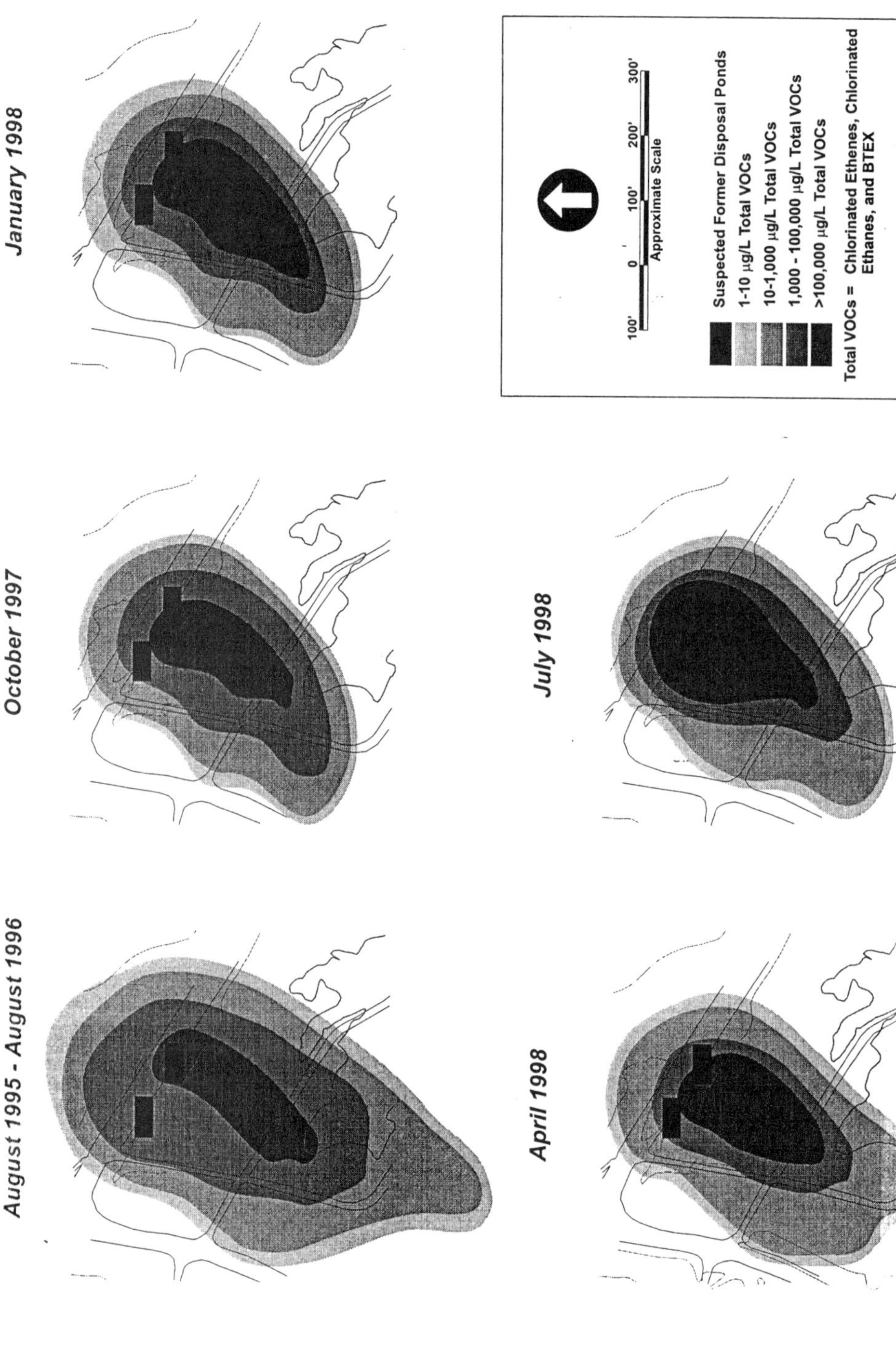

FIGURE 2. Extent of VOC plume over time.

defined by the results of each sampling event, shows that the southern boundary of the plume has instead retreated significantly between 1996 and 1998. The fact that the plume has diminished in size rather than expanded suggests that natural attenuation processes are controlling plume migration.

Concentration trends within the plume also provide evidence of natural attenuation. Maximum detected concentrations of parent compounds PCE and TCE declined significantly between August 1996 and July 1998, as shown in Table 1. Maximum detected concentrations of *cis*-1,2-DCE and VC have increased within the plume. However, in wells farther downgradient from the source area, concentrations of DCE and VC detected in August 1996 (up to 110 micrograms per liter [µg/L] and 120 µg/L, respectively) decreased below detection limits by 1998.

TABLE 1. Maximum detected concentrations of major CAHs.

Analyte	August 1996 (µg/L)	October 1997 - July 1998 (µg/L)
PCE	1,200J	ND(<13) – ND(<5,000)
TCE	11,000J	61 – 180J
cis-1,2-DCE	130,000J	140,000 - 550,000J
VC	17,000	32,000J – 110,000

ND – Not detected
J – Estimated concentration

The extent of BTEX also has decreased substantially since August 1996. Historical data show that BTEX compounds were prevalent in site groundwater. Data collected from October 1997 through July 1998 indicate detected BTEX only in the three wells closest to (and downgradient from) the suspected source.

Chemical and Geochemical Indicators of Biodegradation. Biodegradation is characterized by the consumption of electron acceptors and the production of metabolic byproducts and daughter products. Groundwater geochemical data for the site demonstrate a strongly reducing environment within the plume that is conducive to the reductive dechlorination of CAHs.

As shown in Table 2, concentrations of the electron acceptors DO, nitrate, and sulfate are depleted within the plume (relative to areas outside the plume), indicating the occurrence of aerobic respiration, nitrate reduction, and sulfate reduction. Concentrations of the metabolic byproducts Fe^{+2}, methane, and carbon dioxide are substantially elevated within the plume, suggesting the occurrence of iron reduction and methanogenesis. Because biodegradation reactions are often exothermic, elevated temperatures within the plume appear to be an additional indicator. ORP values support the other geochemical data, indicating strongly reducing conditions within the plume. Figure 3 shows that the isopleths of the major geochemical parameters have an excellent correlation with the distribution of VOCs in groundwater. The only parameter that did not show a clear correlation with the VOC data was chloride, which was consistently detected at the highest concentrations in the well lying closest to the stormwater slough. The

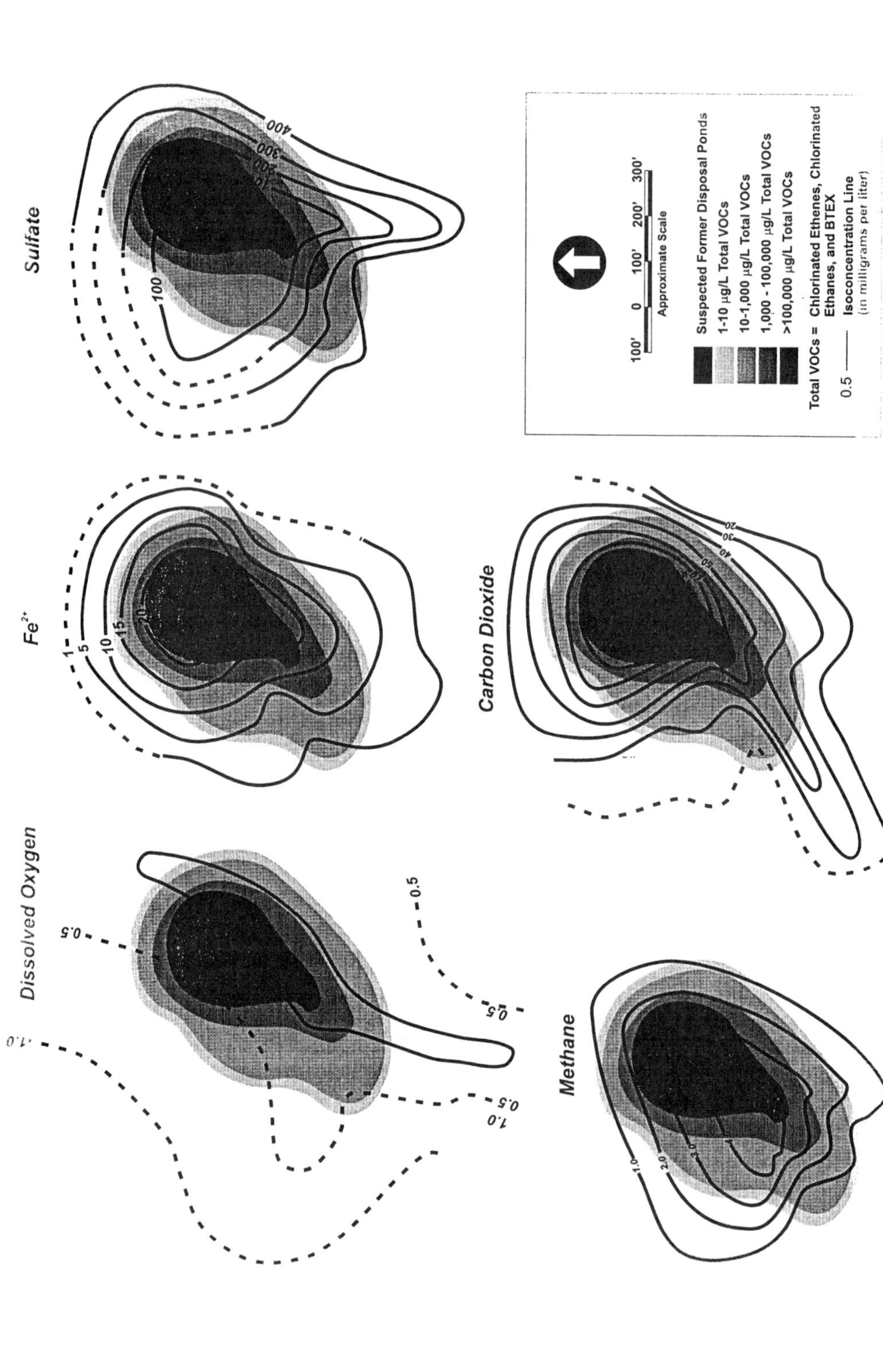

FIGURE 3. Distribution of major geochemical parameters (July 1998).

observed distribution of chloride is interpreted to be influenced by the site's proximity to the ocean.

TABLE 2. Comparison of geochemical parameters inside and outside plume (July 1998).

Parameter	Units	Source Area	Inside Plume, Downgradient of Source Area	Background
Daughter Products				
Ethene	µg/L	210	23,000	<0.50
Ethane	µg/L	2.4	37	<0.50
Electron Acceptors				
Dissolved Oxygen	mg/L	0.28	0.08	1.07
Nitrate	mg/L	<0.20	<0.20	12
Sulfate	mg/L	<0.80	<0.80	590
Metabolic Byproducts				
Ferrous Iron	mg/L	33.33	15.09	0.04
Methane	mg/L	4.2	3.7	0.00025
Carbon Dioxide	mg/L	>100	55	15
Environmental Indicators				
Temperature	°C	23.4	23.4	21.1
Alkalinity	mg/L	660	980	380
ORP	mV	-109.1	-108.8	96.9

In addition to the aforementioned parameters, concentrations of H_2 are used to evaluate oxidation/reduction processes in groundwater because nitrate-, iron-, sulfate-, and carbon-dioxide-reducing (methanogenic) microorganisms exhibit different efficiencies in utilizing the hydrogen in groundwater (Wiedemeier *et al.*, 1998). Because each terminal electron-accepting process has a characteristic H_2 concentration associated with it, H_2 concentrations can be an indication of predominant redox processes. These characteristic ranges are given in Table 3.

TABLE 3. Range of hydrogen concentrations for a given terminal electron-accepting process.

Terminal Electron-Accepting Process	Hydrogen (H_2) Concentration (nm/L)
Denitrification	<0.1
Iron Reduction	0.2 to 0.8
Sulfate Reduction	1 to 4
Reductive Dechlorination	>1
Methanogenesis	5 to 20

Source: Wiedemeier *et al.*, 1998.

Dissolved hydrogen data collected in October 1997 from wells within the VOC plume ranged from 1.40 nanometers per liter (nm/L) to 24.86 nm/L, well above the minimum concentration characteristic of reductive dechlorination. The concentrations of ethene and ethane (i.e., the final daughter products of chlorinated-ethene reductive dechlorination) detected within the plume support the H_2 data (Table 2). Because the only practical source of ethene and ethane is the reductive dechlorination process, the presence of these compounds at such significant concentrations near and downgradient from the source area provides conclusive evidence that CAHs in site groundwater are degrading to innocuous end products via reductive dechlorination.

Along the downgradient edge of the plume, geochemical data indicate oxidizing conditions that may be caused by the recharge of DO through putting-green irrigation. Together with the fact that concentrations of VC historically detected in this vicinity generally have decreased significantly under these oxidizing conditions, the data suggest that destruction of VC via aerobic biodegradation may be occurring in the downgradient portion of the plume.

SUMMARY AND CONCLUSIONS

Abundant evidence exists to document that groundwater remediation through natural attenuation, primarily by biodegradation, is occurring at the site. The lines of evidence indicate that the CAH plume exhibits mixed behavior, with anthropogenic carbon (BTEX) driving the reductive dechlorination of chlorinated solvents throughout the majority of the plume and more aerobic conditions causing VC destruction in a downgradient portion of the plume. It was demonstrated that these processes are reducing the concentration of CAHs and BTEX dissolved in site groundwater, and are limiting plume migration to a degree sufficient to ensure that groundwater protection standards are met at the stormwater slough downgradient from the plume. Continued monitoring will provide additional evidence to ensure that natural attenuation processes will continue to protect human health and the environment.

REFERENCES

National Research Council. 1993. *In Situ Bioremediation - When Does It Work?* National Academy Press, Washington DC.

United States Environmental Protection Agency (USEPA). 1997. *Use of Monitored Natural Attenuation at Superfund, RCRA Corrective Action, and Underground Storage Tank Sites.* Office of Solid Waste and Emergency Response Directive 9200.4-17.

Wiedemeier, T.H., and Chapelle, F.H. 1998. *Technical Guidelines for Evaluating Monitored Natural Attenuation at Naval and Marine Corps Facilities.* Department of the Navy.

Wiedemeier, T.H., Swanson, M.A., Moutoux, D.E., Gordon, E.K., Hansen, J.E., Haas, P., Wilson, J.T., Kampbell, D.H., and Chapelle, F.H. 1998. *Technical Protocol for Evaluating Natural Attenuation of Chlorinated Solvents in Ground Water.* EPA/600/R-98/128, Office of Research and Development, Washington DC.

MODELING CHLOROETHENE BEHAVIOR IN GROUNDWATER IN SUPPORT OF NATURAL ATTENUATION

Andrew Barton, Mark Kelley (Battelle Memorial Institute, Columbus, Ohio)
Art Fisher (NAS Fallon, Nevada)

ABSTRACT: An extensive monitoring program has been implemented in the area surrounding a former Fire Training Pit (FTP) at Naval Air Station (NAS) Fallon, Nevada, for the purpose of evaluating intrinsic remediation of chloroethene contamination in the soil and groundwater. The program consisted of semiannual groundwater sampling to determine trends in contaminant distribution in the shallow groundwater, soil sampling in the vadose zone to delineate the suspected source zone, and groundwater modeling to predict the long-term fate and transport of the contaminants. The site characterization program identified tetrachloroethene (PCE) as the primary chloroethene present in the soil. PCE and related daughter products trichloroethene (TCE) and *cis*-1,2-dichloroethene (*c*DCE) are all present in groundwater at concentrations above maximum contaminant levels (MCLs). Groundwater modeling was performed to simulate the existing chloroethene contamination in groundwater and to determine whether the chloroethene plumes will reach the downgradient point of exposure (POE) before concentrations are reduced below MCLs. These results will be used to determine whether natural attenuation is a viable alternative to active remediation for the chloroethene contaminated groundwater. The groundwater modeling indicated that the inclusion of biodegradation in the simulations is necessary to successfully recreate the existing contaminant plumes and to prevent chloroethene contamination from reaching the POE.

INTRODUCTION

NAS Fallon is located in northwestern Nevada near the city of Fallon, approximately 50 miles east of Reno. The FTP at NAS Fallon is located within the area designated as Site 1 and was used to conduct fire training exercises from the mid-1950s until 1988 (a period of approximately 35 years). It is estimated that approximately 1.1 million gallons of flammable liquids were burned in the FTP, 90% of which was fuel, 9% oil, and 1% solvents (Naval Energy and Environmental Support Activity, 1984). As a result of these exercises, soil and groundwater in the vicinity of the FTP is contaminated with fuel-derived compounds (e.g., benzene, toluene, ethylbenzene, xylenes (BTEX), naphthalene) and chloroethenes (PCE, TCE, DCE). Existing chloroethene groundwater plumes are shown in Figure 1.

CONCEPTUAL MODEL

A shallow, unconfined, silty-sand aquifer is present at NAS Fallon that is underlain by a continuous clay confining layer. Depth to water is typically 5-10 ft below ground surface (bgs) with a saturated thickness ranging from 10-15 ft.

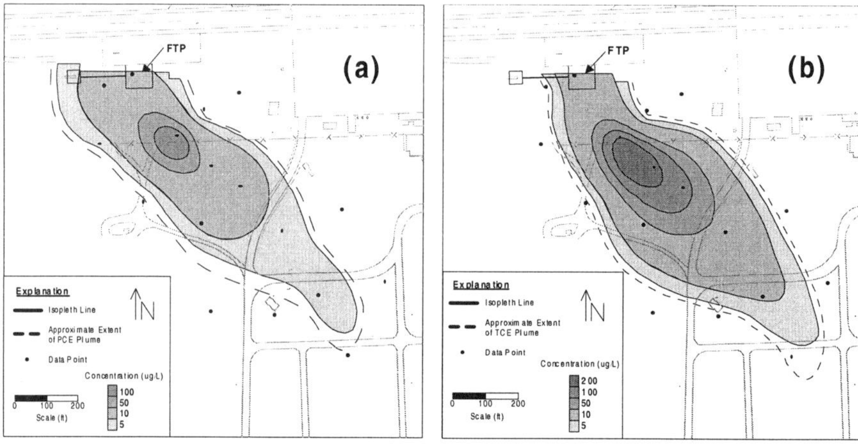

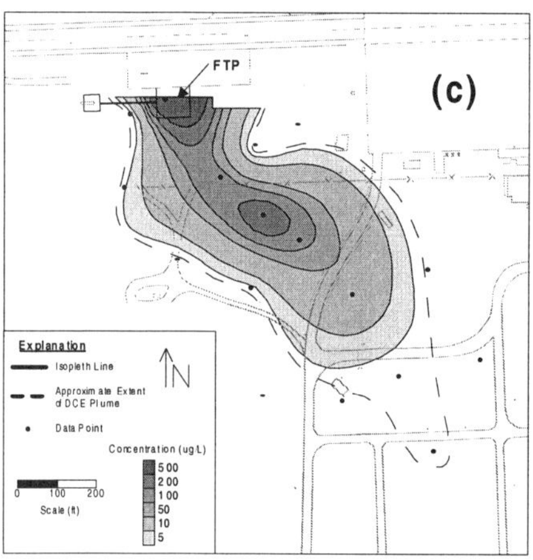

Figure 1. Existing plumes of (a) PCE, (b) TCE, and (c) total DCE at Site 1 (based on data from March/ September 1997)

Groundwater flow is toward the southeast with a hydraulic gradient of approximately 0.001 ft/ft. The average hydraulic conductivity of this shallow aquifer is estimated to be between 10 and 50 ft/d. Seasonal irrigation is practiced in the region, and a partially penetrating drainage ditch (the Lower Diagonal Drain [LDD]) that routes irrigation return flow traverses the model area approximately 2,950 ft downgradient of the FTP. This ditch serves as the POE in the modeling investigation.

Site characterization consisted of five semiannual groundwater sampling events to determine trends in contaminant concentrations throughout the plume of chloroethene contaminated groundwater and to look for evidence of intrinsic biodegradation of the chloroethenes, and soil sampling to quantify the contaminant mass in the source area. PCE appears to have been the primary chloroethene used at the FTP, with average concentrations of approximately 38 mg/kg in the soil underlying the FTP. TCE and *c*DCE were also detected in the soil in the FTP, however, the average concentrations of these compounds were much lower (1.1 and 1.7 mg/kg, respectively) compared to those of PCE. The total mass of PCE, TCE, and *c*DCE in the soil in the vicinity of the source area was estimated to be 17,317 g, 306 g, and 126 g, respectively. A portion of the contaminated soils in the FTP was removed after completion of the source characterization; as a result, an estimated 6,216 g, 158 g, and 50 g of PCE, TCE, and *c*DCE, respectively, remain in the source area.

Although the average concentration of *c*DCE in the soil in the source area is relatively low compared to PCE, it is the most prevalent chloroethene in the groundwater, with a maximum concentration of 3.4 mg/L in the immediate vicinity of the FTP. In contrast, the concentration of PCE and TCE in groundwater in the source area is low (less than 0.05 mg/L). This pattern is attributed to the nearly complete conversion of PCE to *c*DCE in the FTP soil. Maximum concentrations of PCE (0.11 mg/L) and TCE (0.15 mg/L) in groundwater at the site occur approximately 230 ft downgradient of the FTP, suggesting that historical loading rates were higher than current loading rates. This observation is supported by groundwater monitoring data from the immediate vicinity of the FTP, which shows that concentrations of PCE and TCE have declined significantly since 1993. This decline is likely the result of the cessation of FTP activities.

A significant decrease in the concentration of *c*DCE over a short distance downgradient of the FTP suggests that *c*DCE is undergoing degradation in this part of the site. The absence of all but trace levels of vinyl chloride and ethene in this "rapid-reduction" zone suggests that oxidation is the primary pathway for *c*DCE degradation, a process which has been shown to occur under redox conditions similar to those in this portion of the site (Bradley and Chapelle, 1998; Bradley *et al.*, 1998). Downgradient of the rapid-reduction zone, it is unclear whether or not *c*DCE is undergoing further transformation. Because oxidation results in transformation of *c*DCE directly to carbon dioxide without forming daughter products, degradation of *c*DCE in this portion of the site can only be proven by demonstrating a significant decreasing trend in concentrations over time. Similarly, it is unclear whether or not PCE and TCE undergo transformation downgradient of the rapid reduction zone. Although *c*DCE is present in this area, it is believed to have migrated from upgradient rather than forming in place. Therefore, as with *c*DCE, proof of PCE and TCE degradation in this portion of the site requires evidence of a decreasing trend in concentration over time.

Statistical analyses (Mann-Kendall test, linear regression) were performed on monitoring data collected during the intrinsic remediation study, augmented with earlier data collected at the site, and used to investigate the status of the chloroethene plumes. The analyses provide evidence for a decreasing source (even prior to the partial source excavation) and downgradient movement of the plume center of mass. Although no significant trends were detected at the leading edge of the plumes, chloroethene concentrations in the monitoring well at the leading edge of the plume appear to be increasing, suggesting that the plumes may in fact be expanding. Based on this observation, a contaminant transport model was developed to assess the fate of the chloroethene plumes.

NUMERICAL MODELING - RESULTS AND DISCUSSION

A numerical groundwater flow and contaminant transport model was created for the site using the Modular Three-Dimensional Finite-Difference Groundwater Flow Model (MODFLOW) code (McDonald and Harbaugh, 1988) coupled with the Reactive Transport in Three Dimensions (RT3D) code (Clement, 1997). The model was used to simulate the flow and transport of chloroethenes in the shallow groundwater while accounting for effects of natural attenuation processes including dispersion, sorption, and biodegradation.

Calibration of the contaminant transport model involved simulating the formation of existing plumes of PCE, TCE, and total DCE, based on assumed historical release rates and known site conditions. The total DCE plume represents a summation of concentrations of the three DCE isomers; however, *c*DCE comprises the vast majority of total DCE mass. Three scenarios using different transport conditions were conducted during model calibration to help understand the predominant processes active at the site: advection plus dispersion; advection plus dispersion and sorption, and advection plus dispersion, sorption, and biodegradation.

Following model calibration, predictive simulations were performed to investigate the potential extent of future plume migration. Multiple simulations were performed to investigate the effect of different contaminant release scenarios and biodegradation rates. The transport simulations were run until one of the following conditions was met: (1) plume expansion ceased; or (2) a MCL was exceeded at the point of action (POA). The POA is located 150 ft upgradient of the LDD, which the Nevada Department of Environmental Protection (NDEP) considers to be the nearest POE. The POA is located approximately 2,750 ft downgradient of the FTP and 1,650 ft downgradient of the leading edge of the chloroethene plumes.

Model Calibration. The calibrated transport model consisted of nine five-year time steps, each based on steady-state flow conditions represented with the calibrated groundwater flow model. Contaminant mass was released into the model throughout the 45-year period using a set of 12 injection wells with a specified flow rate of 1 ft^3/d for each well. Concentrations of the contaminants injected varied during the nine time steps, but generally were higher during the initial 35 years that the FTP was active, and decreased in the last 10 years following cessation of activities at the FTP. Measured contaminant concentrations from the March and September 1997 sampling events were used as a calibration data set for the transport model. Simulations performed without biodegradation (i.e., using only advection, dispersion, and sorption) were unable to reproduce existing plume configurations, indicating that biodegradation is an important process at the site.

Estimates of biodegradation rates were derived from three sources: BIOCHLOR (Aziz *et al.*, 1998) simulations aimed at matching plume centerline concentrations; analysis of concentration data using the Bucheck and Alcantar (1995) method; and published literature values. Biodegradation rates were varied during the calibration process and were bounded by the derived values. It was necessary to divide the model area into two zones with different biodegradation rates in order to successfully recreate the existing chloroethene plumes. A zone near the former FTP (Zone 1) was modeled using a higher degradation rate for the contaminants than what was specified elsewhere in the model (Zone 2) (Table 1). The area with higher biodegradation rates corresponds to the rapid-reduction zone discussed previously.

Table 1. Biodegradation Rates used in the Calibrated Transport Model

Contaminant	Biodegradation Rate in Zone 1 (d^{-1})	Biodegradation Rate in Zone 2 (d^{-1})
PCE	0.0016	0.00028
TCE	0.0020	0.00036
DCE	0.0025	0.00050

Calibration results indicated a good match between the observed and predicted chloroethene plumes. Predicted concentrations are within an average error of 0.01 mg/L of measured values for PCE, TCE, and DCE. Estimates of actual vs. predicted mass were within 15% for PCE and TCE, and 75% (overestimate) for DCE. DCE values proved to be difficult to simulate in the vicinity of the source area, where a significant decrease in concentration occurs over a very short distance.

A sensitivity analysis was performed on the calibrated transport model to quantify the importance of select input parameters. The analysis indicated that predicted concentrations and plume length are most sensitive to variations in biodegradation rates and source loading rates. In particular, the rate of biodegradation in Zone 2 was determined to be an important factor controlling the extent of plume migration.

Predictive Modeling Results. Predictive simulations were designed to account for the sensitivity of model results to biodegradation rates and the uncertainty in this parameter. Predictive contaminant transport scenarios included:

(1) transport under current conditions (i.e. conditions used to simulate the existing plumes were held constant into the future);
(2) transport involving no biodegradation at the site in the future; and,
(3) transport to determine the minimum biodegradation rate required to prevent contaminant concentrations from exceeding MCLs at the POA.

For each biodegradation scenario, two simulations were performed involving different release scenarios, including: (1) no future source loading (i.e., source removal); and, (2) finite source loading for a period of time necessary to deplete the remaining source mass. The source term in the finite loading simulations was represented by a set of four injection wells covering the portion of the source area remaining at the site following excavation activities in 1999. All future groundwater contamination was assumed to result from the transformation of residual PCE in the soil. PCE mass was depleted from the source area using a half-life of approximately 20 years until mass was no longer present (<1 g) at the site (approximately 260 yrs). TCE and DCE mass loading rates were also assumed to decrease exponentially starting with the rates specified in the last time step of the calibrated model. Residual PCE in the source decreased over time due to loading of PCE, TCE, and DCE.

In the first biodegradation scenario, all three chloroethene plumes are predicted to degrade within 65 years and to stabilize before reaching the POA (Table 2). The same result was obtained using both source release assumptions because the predicted source life is short. The PCE, TCE, and DCE plumes reach a maximum downgradient distance of approximately 1,509 ft, 1,606 ft, and 1,669 ft, respectively, from the source area (Figure 2, Table 2). This scenario assumes that biodegradation will continue indefinitely at the specified rates. However, even if biodegradation in Zone 1 ceases before the source is completely depleted, the plumes will still stabilize upgradient of the POA, provided that the biodegradation continues in the downgradient portion of the site (i.e., Zone 2) at the rate specified in the calibrated model because source concentrations are decreasing (i.e., future concentrations in groundwater will not exceed the maximum existing concentrations at the site). The results of this scenario indicate that the biodegradation rate in the downgradient portion of the site (Zone 2) is the most important factor controlling the extent of plume migration in the future. Unfortunately, there is not conclusive field evidence for biodegradation in this portion of the site (the assumption of biodegradation is primarily based on model calibration).

Table 2. Summary of Transport Simulations Using Finite Source

Transport Scenario	Contaminant	Max. Distance Traveled (ft)	Distance from POA (ft)	Time to Degrade Plume (yr)	Time to Reach POA (yr)	Time to reach POA at MCL (yr)
Calibrated Biodeg.	PCE	1,509	1,241	65	NA	NA
	TCE	1,606	1,144	65	NA	NA
	DCE	1,669	1,081	65	NA	NA
No Biodeg.	PCE	POA	NA	NA	100	120
	TCE	POA	NA	NA	75	85
	DCE	POA	NA	NA	65	80
Critical Biodeg.	PCE	2,527	223	150	NA	NA
	TCE	2,646	104	145	NA	NA
	DCE	2,668	82	180	NA	NA

In the second biodegradation scenario, PCE, TCE, and DCE reach the POA (Figure 3), indicating that biodegradation is necessary in Zone 2 in order for natural attenuation to be viable as a groundwater remedy for this site. In each simulation, PCE, TCE, and DCE reach the POA at 0.001 mg/L after a period of 100 yrs, 75 yrs, and 65 yrs, respectively (Table 2). PCE, TCE, and DCE concentrations are predicted to exceed MCLs at the POA after 120 yrs, 85 yrs, and 80 yrs, respectively (Table 2).

The third biodegradation scenario was conducted to determine the critical (minimum) biodegradation rate necessary to prevent the chloroethenes from reaching the POA at concentrations >0.001 mg/L. The simulations performed for

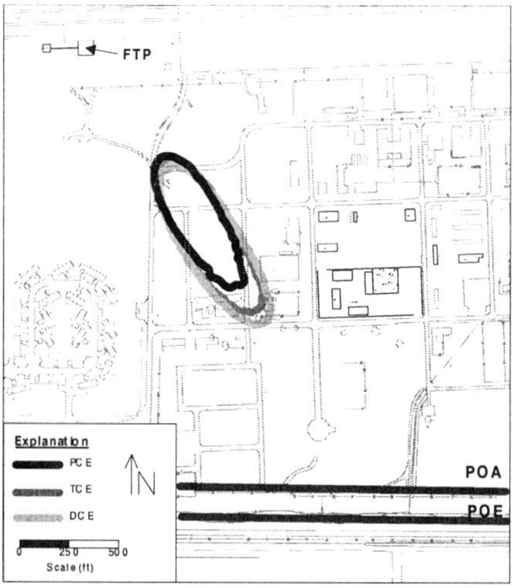

Figure 2. Maximum downgradient extent (>0.001 mg/L) of PCE, TCE, and DCE for biodegradation scenario #1.

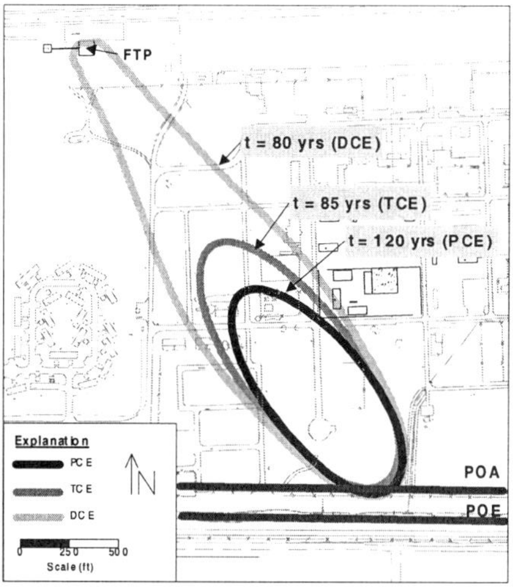

Figure 3. Predicted arrival times at the POA for PCE (>0.005 mg/L), TCE (>0.005 mg/L), and DCE (>0.007 mg/L) for biodegradation scenario #2.

this scenario differed from those in the first scenario because a uniform biodegradation rate was assumed for the entire model area. The simulations indicated that for PCE, TCE, and DCE, a uniform biodegradation rate of 0.00010 d^{-1}, 0.00016 d^{-1}, and 0.00022 d^{-1}, respectively, is sufficient to fully degrade the contaminants before reaching the POA. These biodegradation rates are very low and approach the minimum reported values in the literature for these chemicals. Figure 4 shows that at these rates, the PCE, TCE, and DCE plumes reach a maximum downgradient distance of 2,527 ft, 2,646 ft, and 2,668 ft, respectively (Table 2), but do not impact the POA.

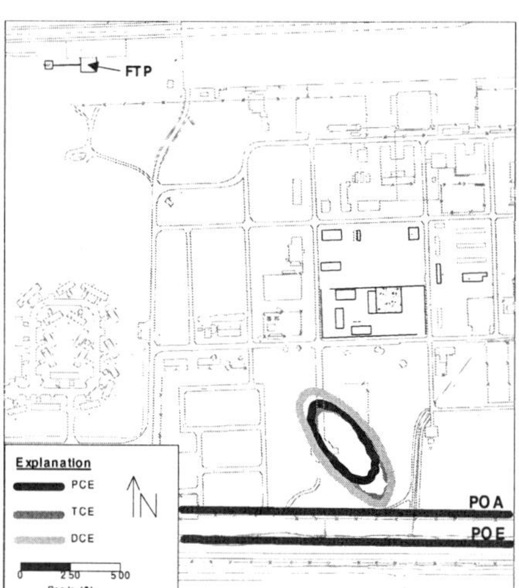

Figure 4. Maximum downgradient extent (>0.001 mg/L) of PCE, TCE, and DCE for biodegradation scenario #3.

CONCLUSIONS

The monitoring program implemented at NAS Fallon was used to determine whether natural attenuation is a viable remedial alternative for groundwater contaminated with chloroethenes. Soil and groundwater sampling provide evidence that source contributions were greater in the past and are decreasing due to a cessation of FTP activities and partial source removal. Based upon the calibrated contaminant transport model, it is evident that a certain amount of biodegradation has taken place at the site. If predicted biodegradation rates remain constant in the future, the chloroethene plumes will fully degrade before reaching the POE. This same result is likely to be true even if biodegradation is occurring at a lower rate than what was derived through model calibration (as long as it exceeds the minimum critical rate determined in this study). However, biodegradation in the downgradient portion of the site (i.e., Zone 2) is uncertain. Concentrations of PCE, TCE, and *c*DCE in this area currently exceed MCLs. If biodegradation does not occur in this area, the contaminants are predicted to reach the POA above their respective MCL regardless of whether or not the source is removed.

REFERENCES

Aziz, C.E., C.J. Newell, J.R. Gonzales, P. Haas, T.P. Clement, and Y. Sun. 1998. *BIOCHLOR Natural Attenuation Decision Support System User's Manual.* Version 1.0. August.

Bradley, P.M. and F.H. Chappelle. 1998. "Microbial Mineralization of VC and DCE Under Different Terminal Electron Accepting Conditions". *Anaerobe* 4: 81-87.

Bradley, P.M., J.E. Landmeyer, and R.S. Dinicola. 1998. "Anaerobic Oxidation of [1,2-^{14}C] Dichloroethene Under Mn(IV)-Reducing Conditions". *Applied Environmental Microbiology* 64: 1560-1562.

Bucheck, T.E. and C.M. Alcantar. 1995. "Regression Techniques and Analytical Solutions to Demonstrate Intrinsic Bioremediation". *In Proceedings of the 1999 Battelle International Conference on In-Situ and On Site Bioreclamation.* April.

Clement, T.P. 1997. *RT3D-Modular Computer Code for Simulating Reactive Multi-Species Reactive Transport in Groundwater Aquifers.* Pacific Northwest National Laboratory, Richland, WA. PNNL-11720.

McDonald, M.G. and A.W. Harbaugh. 1988. *A Modular Three-Dimensional Finite-Difference Ground-Water Flow Model.* Techniques of Water-Resources Investigations 06-A1, U.S. Geological Survey Open-File Report 83-875, 576 p.

Naval Energy and Environmental Support Activity, 1984. *Naval Installation Restoration Program Management Guidance.* NEESA 20.2-047B. Defense Technical Information Center, Port Hueneme, CA.

EVALUATION OF NATURAL ATTENUATION IN A CHLORINATED ETHENE PLUME UNDERGOING GROUNDWATER EXTRACTION

Leo Lehmicke, Christina Noftsker (Exponent, Bellevue, WA), Paul Johansen, Paul Thompson (The Boeing Company, Seattle, WA)

ABSTRACT: Monitored natural attenuation (MNA) is being considered as a replacement for groundwater extraction and treatment as the long-term remedy for a chlorinated ethene plume at the Western Processing Superfund site in Kent, Washington.

The protocol for evaluating natural attenuation of chlorinated solvents suggests collecting at least two lines of evidence: 1) loss of contaminant mass in groundwater over time, and 2) geochemical and biochemical evidence of transformation and degradation. Historical groundwater data were reviewed for evidence of contaminant mass loss over and above that attributable to groundwater extraction. The absence of trichloroethylene (TCE) and the presence of the dechlorination daughter products *cis*-1,2-dichloroethylene (*cis*-DCE) and vinyl chloride indicated TCE was undergoing significant natural biodegradation. A subsequent field investigation provided compelling evidence of natural attenuation. Conditions conducive to reductive dechlorination and anaerobic oxidation were observed in most wells: low dissolved oxygen (DO) and redox, and high concentrations of methane and dissolved iron. Complete dechlorination products (ethene, ethane) were found at levels up to 89 times higher than vinyl chloride on a molar basis.

Modeling plume migration using BIOCHLOR® required estimates of half-lives for *cis*-DCE and vinyl chloride. A half-life for *cis*-DCE of 1.1 years was estimated using data from wells outside the influence of the groundwater extraction system. A half-life for vinyl chloride of 0.55 year was estimated by the same method and two other methods. Using conservative input parameters, modeling indicated vinyl chloride would not migrate beyond the current monitoring well network if groundwater extraction ceased. A monitoring plan that included use of the groundwater extraction system as the contingency remedy was presented to EPA. Tentative approval has been given to shut down the groundwater extraction system in May 2000.

INTRODUCTION

Natural attenuation is evaluated using a "lines of evidence" approach. The three lines of evidence in EPA's *Technical Protocol for Natural Attenuation of Chlorinated Solvents in Groundwater* (U.S. EPA 1998), as taken from the OSWER Directive, are: 1) documented loss of contaminant mass over time, 2) hydrogeological and geochemical data that document transformation or degradation processes and rates, and 3) direct observations of loss in laboratory or field microcosm studies. The third line is required only when the first two lines are insufficient to establish the occurrence of natural attenuation.

BACKGROUND

The 13-acre (5.2-hectare) site is a former chemical recycling facility that operated from 1961 to 1982. Recycled materials and wastes were stored in tanks, drums, and inadequately lined ponds. The shallow (Zone A) and intermediate (Zone B) water-bearing zones are separated by a discontinuous silty clay aquitard. The saturated unit of interest (Zone B) is a silt, sand aquifer with groundwater flow toward the west-northwest at a velocity of approximately 90 ft/year (28 m/year). The plume extends approximately 700 ft west-northwest of the site boundary. It received the misnomer the "trans plume" in 1987 because all DCE was routinely identified by analytical laboratories as the trans isomer. A groundwater extraction and treatment system for the trans plume was installed in 1988, and, as shown in Figure 1, currently consists of three extraction wells (wells 15T2B, 15T3B, and 15T4B) totaling 40 gpm.

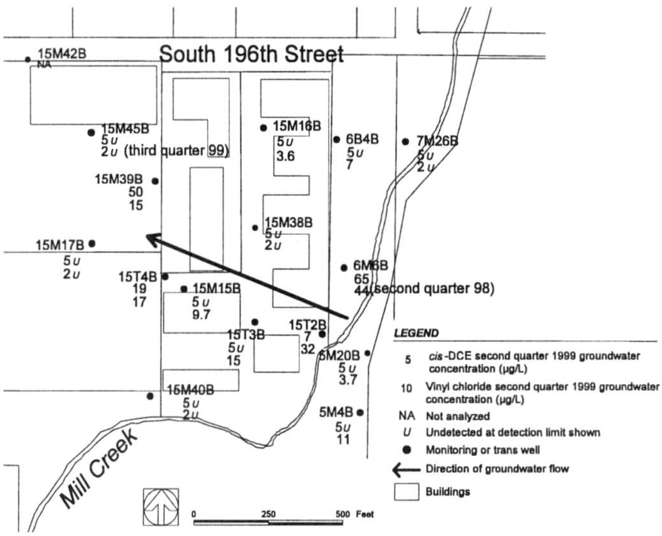

FIGURE 1. Natural attenuation evaluation, monitoring, and performance wells

Natural attenuation was evaluated by reviewing historical groundwater data and collecting new data specific to natural attenuation of chlorinated solvents. Results supporting the first two lines of evidence are discussed below, along with the BIOCHLOR® modeling results, which predict the potential for plume migration if groundwater extraction ceases.

RESULTS AND DISCUSSION

Historical groundwater monitoring data from 1988 through 1998 and geochemical data from the second and third quarters of 1999 were reviewed to more thoroughly evaluate the first two lines of evidence.

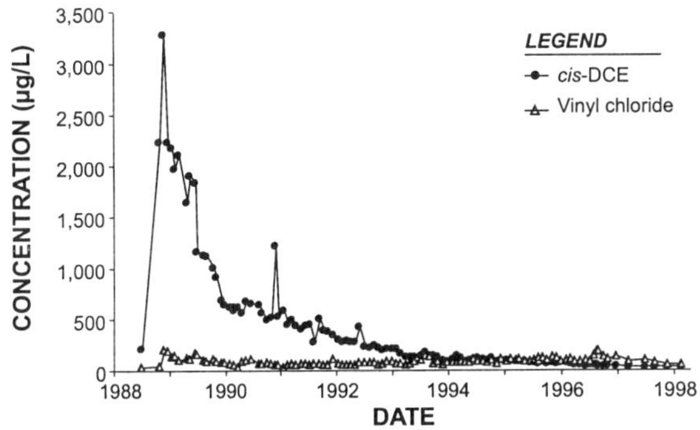

FIGURE 2. ***cis*-DCE and vinyl chloride in Well 15T3B**

First Line of Evidence: Mass Loss. As shown in Figure 2, the pattern of mass loss in an extraction well within the plume shows an initial large decrease in contaminant concentrations due to physical removal, followed by a leveling off at concentrations of approximately 100 μg/L. A different pattern of mass loss is observed in wells outside the influence of the extraction system. These wells exhibit a pattern indicative of natural attenuation due to biological removal as well as an increasing vinyl chloride/*cis*-DCE ratio, as shown in Figure 3. Similar changes in the vinyl chloride/*cis*-DCE ratio are observed in extraction wells (Figure 2), suggesting contaminant removal throughout the plume is a result of both engineered and biological processes. Historical data imply the first line of evidence is satisfied.

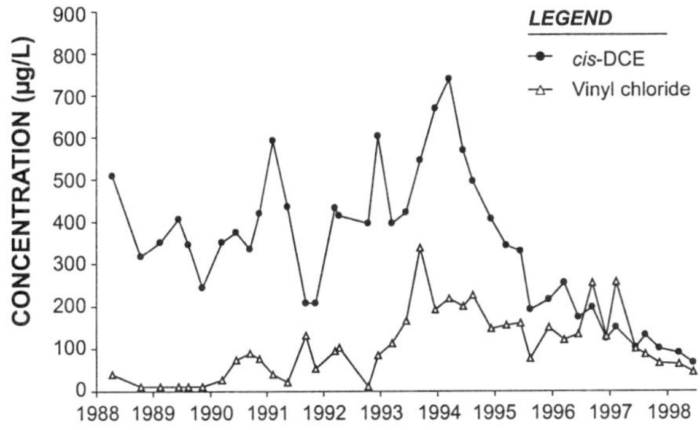

FIGURE 3. ***cis*-DCE and vinyl chloride in Well 6M6B**

Second Line of Evidence: Hydrogeologic and Geochemical Data. Field and geochemical data, including redox potential, DO, methane, ethene, ethane, and dissolved iron concentrations, were collected during two quarterly monitoring events in 1999. Geochemical data that support the second line of evidence are discussed below, including the presence, ratio, and distribution of biodegradation daughter products and geochemical indicators confirming the conditions necessary for the proposed processes. Biological decay rates are discussed in the modeling section.

Biodegradation Daughter Products. TCE undergoes reductive dechlorination under strongly anaerobic conditions and is sequentially transformed to *cis*-DCE, vinyl chloride, ethene, sometimes ethane, and carbon dioxide (McCarty 1994). *cis*-DCE and vinyl chloride have been observed throughout the plume since 1988. The vinyl chloride/*cis*-DCE ratio in well T3 increased from 0.05 in 1988 to more than 1 in 1998. Ethene and ethane were detected in all wells sampled in the second quarter 1999. The (ethene + ethane)/ vinyl chloride ratio ranged from 2 to 89 on a molar basis, as shown in Table 1. The presence of daughter products, the increase in the vinyl chloride/*cis*-DCE ratio, and the high ratio of ethene and ethane to vinyl chloride indicate a large proportion of *cis*-DCE and vinyl chloride have undergone complete reductive dechlorination, as opposed to being removed by the extraction system.

TABLE 1. Molar ratios of (ethene + ethane):vinyl chloride for wells located in trans plume area at Western Processing

Well	Date	Ethene (μg/L)	Ethane (μg/L)	Vinyl Chloride (μg/L)	(ethene + ethane): Vinyl Chloride Molar Ratio
5M004B	04/14/99	12	1	11	2.6
6B004B	04/14/99	39	1	7	12.3
7M026B	04/14/99	1	24	2 *U*	27.0[a]
15M015B	04/14/99	33	28	9.7	13.6
15M016B	04/14/99	104	44	3.6	88.8
15M017B	04/14/99	6	6	2 *U*	13.0[a]
15M039B	04/13/99	15	16	15	4.5
15T002B	04/13/99	36	7	32	2.9
15T003B	04/13/99	12	2	15	2.0
15T004B	04/13/99	12	9	17	2.7

Note: *U* - not detected at detection limit shown

[a] Ratios are based on vinyl chloride detection limit; therefore, ratios are conservative.

Geochemical Indicators. Geochemical indicators confirm that anaerobic conditions conducive to reductive dechlorination are clearly present in Zone B groundwater. Indicators of anaerobic conditions include the absence of detectable DO in 7 of 10 wells, redox values less than −100 mV in 9 of 10 wells, and methane concentrations ranging from 11 to 59 mg/L, with 7 wells above saturation. Vinyl chloride can also undergo anaerobic oxidation under iron-reducing conditions, a process potentially more rapid than reductive dechlorination (Bradley and Chapelle 1996). Dissolved iron was observed at concentrations up to 24 mg/L, indicating conditions conducive to anaerobic oxidation of vinyl chloride.

Third Line of Evidence: Laboratory Microcosm Studies. The EPA protocol does not require a third line of evidence unless data to support the first two lines of evidence are inadequate or inconclusive. Data collected to support the first two lines of evidence were sufficiently strong that conducting laboratory studies was not warranted.

Prediction of Plume Stability Based on Modeling with BIOCHLOR®. BIOCHLOR® was used to predict potential plume migration after cessation of groundwater extraction. BIOCHLOR® is a natural attenuation screening model that can handle the multiple reactions necessary to model reductive dechlorination of chlorinated solvents (Aziz et al. 1999). Modeling using BIOCHLOR® required estimates for biological decay rates (half-lives) for *cis*-DCE and vinyl chloride.

Calculating Biological Half-Lives. Biodegradation half-lives could not be calculated from field data using normal methods because the flow path at the site has been disrupted by groundwater extraction, and a conservative tracer is not available. Half-lives for *cis*-DCE and vinyl chloride were estimated by evaluating historical concentrations in wells outside the influence of the extraction system. In addition, half-lives for vinyl chloride were estimated by two other methods: 1) evaluating the ratio of (ethene + ethane)/vinyl chloride, and 2) confirming that method by calibrating BIOCHLOR® to generate a half-life that fits the ratio of ethene/vinyl chloride observed throughout the plume.

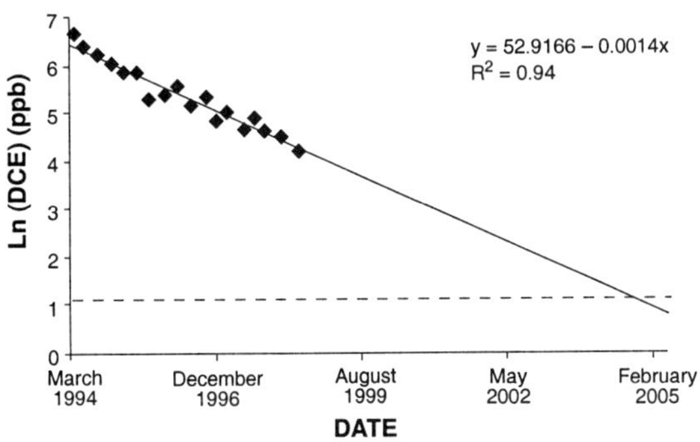

FIGURE 4. Example data used to estimate a half-life

A half-life for *cis*-DCE was estimated from historical concentrations in wells 6M6B, 7M26B, and 15M39B. A half-life can be calculated if concentration data are linear when plotted against time on a log scale (Buscheck et al. 1993). Plots of the natural log of *cis*-DCE concentration vs. time for these wells generated R-squared values between 0.87 and 0.94 (Figure 4). Half-lives were then calculated from the following equation:

$$T_{1/2} = \frac{0.693t}{\log C_t/C_0}$$

where:

$T_{1/2}$ = half-life for *cis*-DCE

t = time between C_0 measurement and C_t measurement

C_0 = initial *cis*-DCE concentration

C_t = final *cis*-DCE concentration.

Results from the three wells are shown in Table 2. The estimated mean half-life for *cis*-DCE used in modeling is 1.1 years.

TABLE 2. Estimated half-life for *cis*-DCE

Well	C_0	C_t	Measurement Period	t (years)	$T_{1/2}$ (years)
6M6B	735	65	3/94–6/98	4.25	1.2
7M26B	158	7	8/90–9/96	6.1	1.4
15M39B	426	50	4/97–4/99	2	0.6
				Mean	1.1

Note: *cis*-DCE - *cis*-1,2-dichloroethene

An estimate for a half-life for vinyl chloride was obtained from historical concentrations in Well 6M6B and generated an apparent half-life of 0.60 year, roughly half the *cis*-DCE half-life. However, if the only mechanism of *cis*-DCE loss is reductive dechlorination to vinyl chloride, then vinyl chloride should be created from *cis*-DCE at the rate of *cis*-DCE loss. The actual half-life for vinyl chloride is a combination of its rate of production and loss (combination of the half-lives) or 0.37 year. Two additional estimates for a half-life for vinyl chloride to ethene were calculated using the two alternative methods described below.

The second method assumes the pumped plume is in a biochemical "steady-state" and the steady-state ratio of (ethene + ethane)/vinyl chloride is proportional to their rates of production. Molar ratios for (ethene + ethane)/vinyl chloride in plume wells ranged from 2 to 89 (Table 1). However, some wells have no *cis*-DCE, and ethene and ethane may accumulate if there is no other loss mechanism. The most appropriate steady-state ratios would be observed in wells that contain *cis*-DCE and have a physical removal mechanism (wells near the point of extraction). The ratios in these wells are at the lower end of the range (2.7 to 4.5) and suggest that the *cis*-DCE half-life is three to five times longer than the vinyl chloride half-life.

The third method involved calibrating BIOCHLOR® to extrapolate a half-life that gave ratios of ethene/vinyl chloride similar to those observed in the plume, and is, therefore, a confirmation of the second method. BIOCHLOR® was run to steady-state using current conditions at Well 15M39B. Ratios of ethene/vinyl chloride were not calculated for the source point, the first downgradient point, or distant downgradient points where the concentration of vinyl chloride or ethene dropped below 1 μg/L. Table 3 shows ratios of ethene/vinyl chloride obtained from BIOCHLOR® simulations using different pairs of half-lives for vinyl chloride and *cis*-DCE and the actual (ethene + ethane)/vinyl chloride ratios observed in plume wells, which range from 2 to 89. A similar range of ratios (2.3–60) is predicted by a BIOCHLOR® simulation using a half-life for *cis*-DCE of three times the half-life for vinyl chloride. The three methods suggest the half-life for *cis*-DCE is at least twice as long as the half-life for vinyl chloride. The half-lives used in the BIOCHLOR® modeling exercise were 1.1 years and 0.55 year for *cis*-DCE and vinyl chloride, respectively, and are consistent with values in the literature. Half-lives in the draft and final versions of the EPA protocol (U.S. EPA 1998) range from 0.21 year to 3.8 years for *cis*-DCE, and 0.24 year to 4.6 years for vinyl chloride, with a mean of 1.2 years.

TABLE 3. Range of ethene:vinyl chloride ratios estimated from BIOCHLOR® modeling and observed in actual plume, second quarter 1999

Simulation/ Plume	*cis*-DCE Half-life (years)	Vinyl Chloride Half-life (years)	*cis*-DCE:Vinyl Chloride Half-life Ratio	Ethene:Vinyl Chloride Ratio
BIOCHLOR®	1.1	2.2	0.5	0.8–4.3
BIOCHLOR®	1.1	1.11	1.0	1.1–10.5
BIOCHLOR®	1.1	0.55	2.0	1.8–30
BIOCHLOR®	1.1	0.37	3.0	2.3–60
Actual Plume				2.0–89

Modeling Results and Conclusions. Modeling using data from the second and third quarters of 1999 predicted the vinyl chloride plume could migrate a maximum of 390 to 460 ft (120–142 m), respectively, downgradient from Well 15M39B, but would still be within the current monitoring well network. Predicted migration distances are likely further than what will actually occur because of the use of generally conservative input parameters in this modeling exercise.

The EPA OSWER directive is a clarification of EPA's policy regarding the use of MNA at Superfund sites (U.S. EPA 1999). The key elements of concern are: adequate site characterization, use of MNA in conjunction with other remedial measures (specifically source control), plume stability, a reasonable time frame compared to other remedies, adequate performance monitoring, and contingency remedies. Historical remedial actions in combination with plume stability predictions and cleanup timeframe indicate all the elements of concern can be met for the site. A monitoring plan that included use of the groundwater extraction system as the contingency remedy was presented to EPA. Tentative approval has been given to shut down the groundwater extraction system in May 2000.

REFERENCES

Aziz, E.E., C.J. Newell, J.R. Gonzales, P. Haas, T.P. Clement, and Y. Sun. 1999. BIOCHLOR natural attenuation decision support system user's manual. Groundwater Services, Inc., Houston, TX; Air Force Center for Environmental Excellence, Brooks AFB, San Antonio, TX; Battelle Pacific Northwest National Laboratory, Richland, WA

Bradley, P.M., and F.H. Chapelle. 1996. Anaerobic mineralization of vinyl chloride in Fe(III)-reducing, aquifer sediments. Environ. Sci. Technol. 40:2084–2086.

Buscheck, T., K.T. O'Reilly, and S.N. Nelson. 1993. Evaluation of intrinsic bioremediation at field sites. In: Proc. Petroleum Hydrocarbons and Organic Chemicals in Groundwater: Prevention, Detection and Restoration. Houston, TX.

McCarty, P. L. 1994. An Overview of Anaerobic Transformation of Chlorinated Solvents: In: *Symposium on Intrinsic Bioremediation in Groundwater, Denver, CO, August 30–September 1, 1994*, P. 135-142.

U.S. EPA. 1998. Technical protocol for evaluating natural attenuation of chlorinated solvents in groundwater. EPA/600/R-98/128. Website located at: http://ftp.epa.gov/pub/ada/reports/protocol.pdf, last visited October 1, 1999. U.S. Environmental Protection Agency, Office of Research and Development.

U.S. EPA. 1999. *Use of Monitored Natural Attenuation at Superfund, RCRA Corrective Action, and Underground Storage Tank Sites.* EPA 9200.4–17. U.S. Environmental Protection Agency, Office of Solid Waste and Emergency Response, Washington, DC.

REDUCTIVE DECHLORINATION OF CHLORINATED ALIPHATIC COMPOUNDS AS A RESULT OF A PETROLEUM FUEL RELEASE CASE STUDY OF THE FORMER PHOENIX BUS TERMINAL, PHOENIX, ARIZONA

James N. Clarke, R.G (LAWGibb, Phoenix, Arizona U.S.A)

ABSTRACT: It has been demonstrated that biodegradation of petroleum hydrocarbons in groundwater creates conditions (low dissolved oxygen, low oxidation/reduction potential, and production of hydrogen [H_2]) that are favorable to reductive dechlorination (sequential removal of chlorine atoms and replacement by hydrogen atoms) of chlorinated aliphatic compounds (CACs). The Former Phoenix Bus Terminal (site) is located in downtown Phoenix, Arizona and within an area referred to as the Motorola 52nd Street National Priorities List (NPL) Site. Groundwater within the Motorola 52nd Street NPL Site is impacted by tetrachloroethene (PCE), trichloroethene (TCE), 1,1,1-trichloroethane (TCA), cis-1,2-dichloroethene (c-1,2-DCE), 1,1-DCE, and 1,1-dichloroethane (1,1-DCA). The site is not a source of the CACs; however, releases of an unknown amount of gasoline and diesel fuel have impacted the groundwater below the site. Results of the groundwater monitoring program indicate that reductive dechlorination of the CACs is occurring. Additionally, aerobic biodegradation is containing the down-gradient migration of dissolved benzene, vinyl chloride, and methane. A long-term monitored natural attenuation remedy has been approved for the site and a request for risk-based closure of the site will be made in the near future using the premise that the petroleum plume is stable and is beneficially mitigating the CACs.

INTRODUCTION

Site Description and Conditions. The former Phoenix Bus Terminal was located in downtown Phoenix, Arizona (see Figure 1) (Olm and Ruddiman, 1998) and was operated from 1968 to 1996. The operations included a fuel hydrant system that dispensed diesel fuel and gasoline to buses and other vehicles owned by the bus company. An initial release of fuel hydrocarbons was reported at the site in 1985, which resulted in the installation of 7 monitoring wells at the site. In May 1990, more than 1 ft (0.3 m) of free-phase diesel fuel was observed in Well MW-1 (see Figure 2). This observation initiated subsequent assessment and remedial activities. To date, 40 soil borings have been drilled and sampled, 23 groundwater monitoring wells have been installed, and a biovent system was installed and operated. A parking garage has recently been constructed on the site.

The site is situated within a broad alluvial basin referred to as the Salt River Valley. Approximately 1,500 ft (457 meters) of unconsolidated sediments underlie the site and depth to groundwater ranges seasonally from 60 ft (18 m) below ground surface (bgs) to 90 ft (27.5 m) bgs. The entire sediment thickness

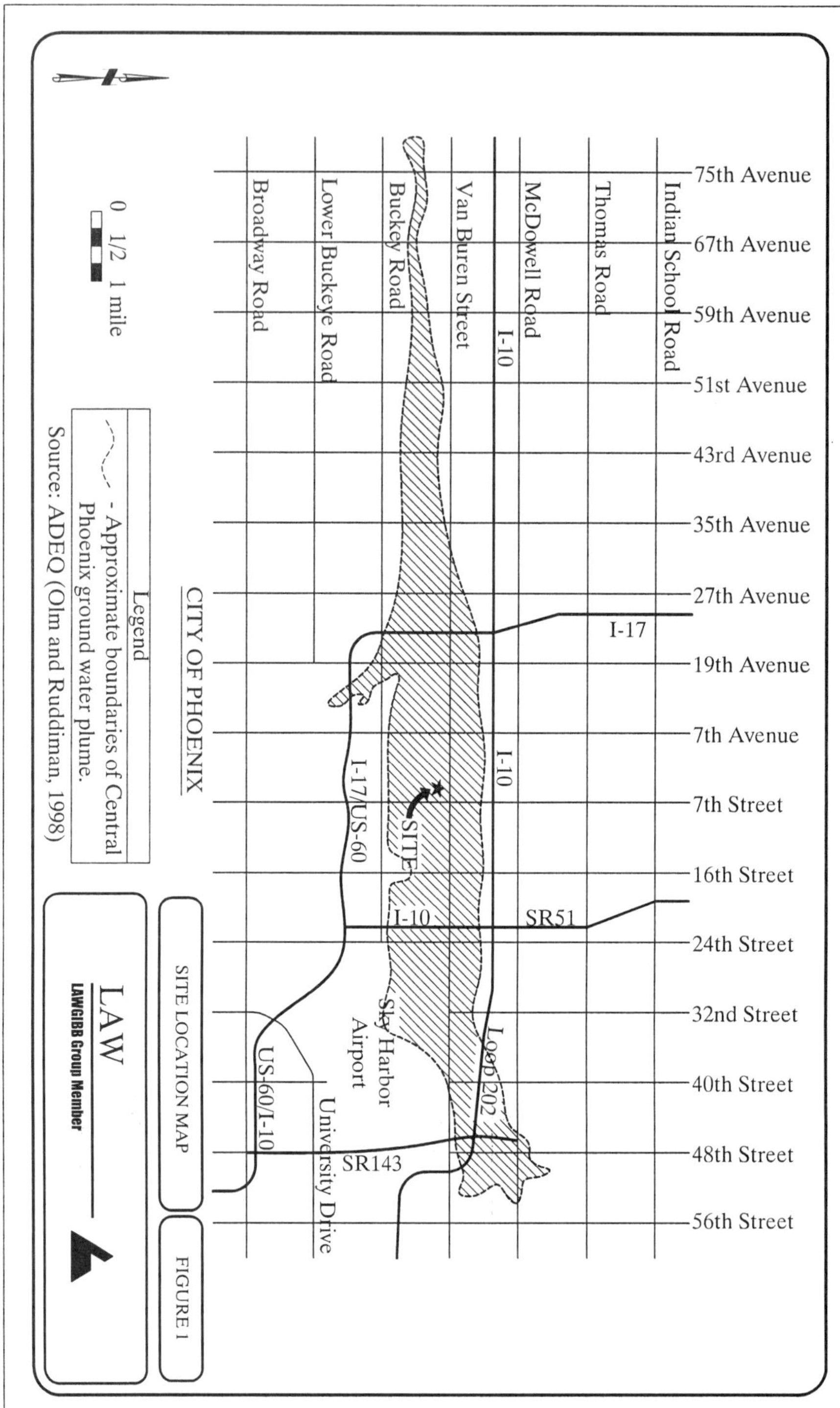

75th Avenue
67th Avenue
59th Avenue
51st Avenue
43rd Avenue
35th Avenue
27th Avenue
19th Avenue
7th Avenue
7th Street
16th Street
24th Street
32nd Street
40th Street
48th Street
56th Street
Indian School Road
Thomas Road
McDowell Road
I-10
Van Buren Street
Buckey Road
Lower Buckeye Road
Broadway Road
I-17
I-10
I-17/US-60
SITE
I-10
SR51
Sky Harbor Airport
Loop 202
US-60/I-10
SR143
University Drive
0 1/2 1 mile
CITY OF PHOENIX
Legend
- Approximate boundaries of Central Phoenix ground water plume.
Source: ADEQ (Olm and Ruddiman, 1998)
SITE LOCATION MAP
FIGURE 1
LAW
LAWGIBB Group Member

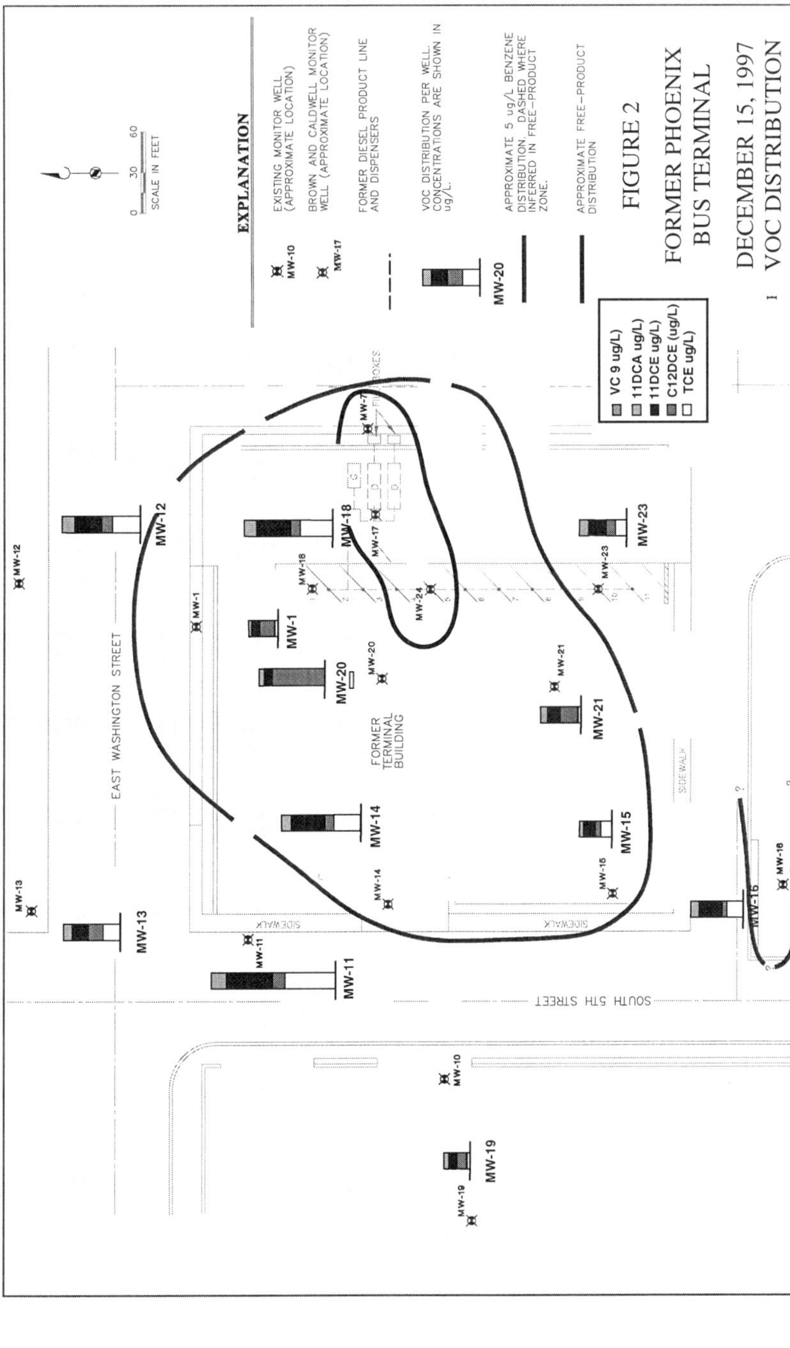

FIGURE 2
FORMER PHOENIX BUS TERMINAL
DECEMBER 15, 1997
VOC DISTRIBUTION

from the water table to bedrock is saturated. Site characterization activities have been conducted to a depth of 120 ft (36.5 m) bgs and the soils below the site are predominantly comprised of highly transmissive sand, gravel, and cobbles. However, from approximately 60 ft (18 m) bgs to 95 ft (29 m) bgs the soils contain 10 to 20 percent clay, which has promoted the formation of a hydrocarbon smear zone.

The groundwater below the site is impacted by benzene, toluene, ethylbenzene, xylene (BTEX), total fuel hydrocarbons (TFH), and free-phase fuels that originated from the site and an adjacent upgradient source. Benzene is the only site-specific constituent that has been detected above the Arizona Aquifer Water Quality Standards (AAWQS) and is the primary compound of concern. The site is also located within an area now known as the Motorola 52nd Street NPL Site and groundwater within this area is known to contain concentrations of PCE, TCE, c-1,2-DCE, TCA, 1,1-DCE and 1,1-DCA. As shown on Figure 1 (Olm et al., 1998), the CAC plume is approximately 2 miles (3.2 km) wide and more than 10 miles (16 km) long. The former bus terminal operator/owner is not a Potentially Responsible Party (PRP) to the known CAC plume.

Objectives. Based on experience and observations at similar sites in the Phoenix area, it was anticipated that biodegradation of the petroleum hydrocarbons was promoting reductive dechlorination of the known CACs in the groundwater below the site. Agricultural and livestock activities were formerly conducted upgradient of the site. Therefore, sufficient nutrients and electron donors, such as nitrates and sulfates, would be expected in the groundwater. Natural attenuation had rarely been proposed as a remedial option in Arizona. Therefore, a well designed data collection and evaluation program was necessary to demonstrate that natural attenuation was occurring and was a feasible remedial option.

In August 1997, an ongoing corrective action groundwater monitoring program began at the site. The groundwater monitoring program is designed to demonstrate the following: (1) natural attenuation via aerobic and anaerobic biodegradation is containing and mitigating dissolved BTEX in the groundwater; (2) the biodegradation of the petroleum hydrocarbons is beneficially reductively dechlorinating the CACs; (3) active remediation is not necessary due to natural attenuation; and, (4) considering 1 through 3 above and the fact the groundwater in the area of the site is not being used and will not likely be used in the near future, that risk-based closure of the site is an acceptable alternative. Although the natural attenuation remedy has been approved for the site, current aquifer protection laws do not provide for risk-based closure of groundwater impacted LUST sites in Arizona. Arizona is currently in the process of adopting Risk-Based Closure Rules for groundwater impacted sites and it is anticipated that these rules will become effective by September 2000. The site is a candidate to be one of the first groundwater-impacted sites to be closed in Arizona using these rules.

MATERIALS AND METHODS

The monitoring well network at the site consists of 15 monitoring wells, identified as wells MW-1, MW-7, MW-11, MW-12, MW-13 through MW-21,

MW-23 and MW-24 on Figure 2. The wells serve the following purposes: Well MW-12 monitors background conditions; Wells MW-7, MW-17, MW-18, and MW-24 are source area wells and are used to monitor/recover free product; Wells MW-1, MW-20, MW-21, and MW-23 are mid-plume wells (downgradient of free product wells); and Wells MW-11, MW-13, MW-14, MW-16, and MW-19 are the Point-of-Compliance wells, with MW-19 being the furthest downgradient well. The wells are approximately 100 ft. (30.5 m) deep and are screened from 60 ft. (18 m) to 100 ft. (30.5 m). Groundwater sampling and analysis was initially performed on a quarterly basis and is currently performed bi-annually.

Sample Collection and Analysis. Groundwater samples are collected using the micropurging approach. A Grundfos Redi-Flo 2 purging and sampling pump is installed at 85 ft (26 m) bgs, which is within the known petroleum smear zone interval and at a depth where the highest dissolved BTEX concentrations have been observed. Prior to sample collection, the well is purged at a slow flow rate, 0.25 gallons per minute (gpm) (0.95 lpm) to 0.5 gpm (1.9 lpm). The water is passed through a cell and pH, temperature, electroconductivity (EC), dissolved oxygen (DO) and oxidation/reduction potential (ORP) are constantly measured using a multi-parameter monitoring system (Horiba, YSI or equivalent). The parameters typically stabilize within five minutes and less than five gallons (19 l) of water are removed from the well. Once the parameters have stabilized, the groundwater sample is collected at a flow rate of 100 ml/min.

Based on the analytical guidelines provided in the *Technical Protocol for Evaluating Natural Attenuation of Chlorinated Solvents in Groundwater* (Weidemeier, et al., 1996), the groundwater samples are analyzed for the following: volatile organic compounds (VOCs) by EPA Method 524 or 8021; nitrate, sulfate, chloride, and alkalinity; ethene, ethane, and methane by Method RSKSOP-175; and, DO and ORP in the field using the multi-parameter monitoring system. Well MW-12 is the background well and data collected from the other wells are compared to the MW-12 data to evaluate if biodegradation and reductive dechlorination are occurring and to evaluate the progress of the system.

RESULTS

The results of the natural attenuation monitoring program are shown on Figure 2 and in Tables 1 and 2. Figure 2 shows the relative changes in CAC concentrations across the site. Table 1 shows the changes in VOC and parameter concentrations along a transect of the petroleum hydrocarbon plume and Table 2 shows changes in VOC and parameter concentrations along a line that connects the six downgradient point-of-compliance wells (Clarke, 1998). The data presented on Figure 2 and in Tables 1 and 2 were collected during groundwater sampling events that were performed during August and December 1997.

TABLE 1. Changes in VOC and Parameter Concentrations along a Transect of the Petroleum Hydrocarbon Plume

Compound	MW-12 (background)	MW-1 (source)	MW-20 (mid-plume)	MW-19 (downgradient)
BTEX	<0.5 μg/L	54 μg/L	141.2 μg/L	1.2 μg/L
PCE	1.6 μg/L	<5.0 μg/L	<5.0 μg/L	<0.5 μg/L
TCE	43 μg/L	1.8 μg/L	<5.0 μg/L	3.8 μg/L
c-1,2-DCE	8.2 μg/L	32 μg/L	74 μg/L	13 μg/L
1,1-DCE	36 μg/L	26 μg/L	16 μg/L	7.5 μg/L
VC	<0.5 μg/L	3.5 μg/L	2.7 μg/L	0.7 μg/L
Ethene	<40 μg/L	<2000 μg/L	<2000 μg/L	<40 μg/L
1,1-DCA	14 μg/L	NA	7.6 μg/L	5.4 μg/L
Ethane	<40 μg/L	<2000 μg/L	<2000 μg/L	<40 μg/L
DO	2.4 mg/L	0.3 mg/L	0.2 mg/L	0.3 mg/L
Nitrate	22 mg/L	<0.5 mg/L	1.4 mg/L	2.8 mg/L
Sulfate	100 mg/L	30 mg/L	3.4 mg/L	44 mg/L
Chloride	220 mg/L	210 mg/L	210 mg/L	190 mg/L
Alkalinity	280 mg/L	410 mg/L	470 mg/L	400 mg/L
Methane	<40 μg/L	13000 μg/L	7200 μg/L	130 μg/L

TABLE 2. Changes in VOC and Parameter Concentrations in Point-of-Compliance Wells

Compound	MW-13 North	MW-11	MW-14	MW-19	MW-15 South
BTEX	<0.5 μg/L	<0.5 μg/L	34.6 μg/L	1.2 μg/L	19.5 μg/L
PCE	0.6 μg/L	2.0 μg/L	1.6 μg/L	<0.5 μg/L	1.2 μg/L
TCE	23 μg/L	62 μg/L	33 μg/L	3.8 μg/L	13 μg/L
c-1,2-DCE	19 μg/L	12 μg/L	9.7 μg/L	13 μg/L	6 μg/L
1,1-DCE	18 μg/L	36 μg/L	36 μg/L	7.5 μg/L	13 μg/L
VC	<0.5 μg/L	<0.5 μg/L	<0.5 μg/L	0.7 μg/L	<0.5 μg/L
Ethene	<40 μg/L	<40 μg/L	<40 μg/L	<40 μg/L	<40 μg/L
1,1-DCA	11 μg/L	17 μg/L	15 μg/L	5.4 μg/L	6.5 μg/L
Ethane	<40 μg/L	<40 μg/L	<40 μg/L	<40 μg/L	<40 μg/L
DO	4.3 mg/L	0.5 mg/L	0.6 mg/L	0.3 mg/L	0.7 mg/L
Nitrate	3.7 mg/L	22 mg/L	26 mg/L	2.8 mg/L	17 mg/L
Sulfate	64 mg/L	89 mg/L	92 mg/L	44 mg/L	93 mg/L
Chloride	210 mg/L	210 mg/L	200 mg/L	190 mg/L	200 mg/L
Alkalinity	380 mg/L	310 mg/L	280 mg/L	400 mg/L	290 mg/L
Methane	310 μg/L	340 μg/L	100 μg/L	130 μg/L	41 μg/L

As shown in Tables 1 and 2, background groundwater is aerobic (2 to 4 mg/L DO) and is nitrate and sulfate rich. As groundwater moves through the petroleum source area, the system becomes anaerobic resulting in decreased TCE, 1,1-DCE, 1,1-DCA, DO, nitrate and sulfate concentrations and increased c-1,2-DCE, alkalinity and methane concentrations. Vinyl chloride has also been periodically observed in wells closest to the petroleum source area (MW-20 and MW-21). Ethene and ethane have not been detected in the site wells. Natural attenuation data has also been collected during March 1998, December 1998 and December 1999 and the trends observed on Figure 2 and in Tables 1 and 2 have not changed since December 1997.

CONCLUSIONS

Based on the data presented in Tables 1 and 2, the following conclusions are presented:

1. Nitrate reducing, sulfate reducing and methanogenic bacteria are all actively degrading petroleum hydrocarbons in the groundwater below the site.
2. Reductive dechlorination of TCE, c-1,2-DCE, 1,1-DCE and 1,1-DCA is occurring within the anaerobic zone, with almost all the TCE transformed to c-1,2-DCE. The periodically observed vinyl chloride is possibly a daughter product of 1,1-DCE or c-1,2-DCE degradation. However, the parent compound of the vinyl chloride cannot be confirmed by the current data.
3. Aerobic bacteria are apparently degrading BTEX, vinyl chloride and methane along the leading edge of the BTEX plume (MW-13, MW-11, MW-14 and MW-15), thus providing plume containment (see Figure 3). Benzene in these wells has not exceeded the AAWQS of 5 μg/L since March 1998.
4. As indicated by anaerobic conditions and indicators, Well MW-19, the furthest downgradient well on the site, is located directly in the flow path from the petroleum source area and is characterized by a nitrate-reducing environment. Benzene in this well has not exceeded the AAWQS of 5 μg/L.
5. Based on the observation that natural processes are mitigating the dissolved petroleum hydrocarbons and are beneficially reductively dechlorinating the CACs, active remediation of the petroleum hydrocarbon impact is not warranted.

The data indicate that there is a beneficial "symbiotic-like" relationship between the petroleum hydrocarbons and the CACs and between the aerobic and anaerobic environments. This relationship has possibly taken some time to develop and has apparently resulted in a stable BTEX plume. Not only could application of active remedial approaches such as pump-and-treat disturb that stability, they would introduce contaminants to the surface requiring handling, disposal or treatment.

Due to the presence of the CACs, it is unlikely the groundwater will be used as a drinking water supply in the near future. This, combined with the above conclusions, makes the site eligible for risk-based closure.

REFERENCES

Olm, R., and Ruddiman, W., 1998. *Central Phoenix Ground Water Plume Overview Map.* Prepared for Arizona Department of Environmental Quality.

Weidemeier, T.H, Swanson, M.A, Moutoux, D.E, Gordon, E.K, Wilson, J.T, Wilson, B.H, Hansen, J.E, Haas, P. and Chapelle, F.H. 1996. *Technical Protocol for Evaluating Natural Attenuation of Chlorinated Solvents in Groundwater.* Prepared for Air Force Center for Environmental Excellence, Technology/Transfer Division, Brooks Air Force Base, San Antonio, Texas

Clarke, J.N. 1998. *Dissolved-Phase Groundwater Corrective Action Plan, Former Phoenix Bus Terminal, 525 East Washington Street, Phoenix, Arizona.* Prepared for Transportation Leasing Company.

MODELING NATURAL ATTENUATION OF CHLORINATED SOLVENT PLUMES

Anna R. Mason[1] and Thomas J. Franz[2], Beatty Franz & Associates, Bolton, ON, Canada; Mark R. Harkness, GE Corporate Research and Development, Schenectady, NY; Michael Figura, Naval Air Engineering Station, Lakehurst, NJ

ABSTRACT: A groundwater flow and contaminant transport model was developed for use in the evaluation of natural attenuation as the preferred remedial alternative at a site contaminated with chlorinated solvents. The reactive transport code RT3D was used to simulate the sequential biodegradation of PCE, TCE, cis-1,2-DCE, vinyl chloride, and ethene through anaerobic dechlorination and the production of chloride from the aerobic biodegradation of these individual compounds. The calibrated model was used to predict the migration of the chlorinated VOC plumes 70 years into the future. Predictive simulations showed that the extent of the area where low concentrations of chlorinated VOCs (1-10 µg/l) are found may increase slightly in the deep aquifer zone while the sources remain active. However, the chlorinated VOC contaminant concentrations decreased steadily due to natural attenuation processes after source depletion.

INTRODUCTION

Two large chlorinated solvent plumes known as the north and south plumes are located in the areas designated as Areas I and J at the Naval Air Engineering Station (NAES) Lakehurst, New Jersey. The contaminants found in the groundwater plumes include PCE, TCE, 1,1,1-TCA, and their degradation products.

Both plumes have migrated off-site and are approaching, or have reached, a downgradient creek. Concentrations of individual VOCs at the downgradient creek are generally around 1 ppb. The highest contaminant concentrations are found in the 'deep' zone of the aquifer, between 55 and 75 ft below ground surface. The size of the plumes, as defined by the 1 ppb contour, ranges from approximately 4000 ft to 6000 ft in length. The plumes are both approximately 2000 ft wide.

A groundwater flow and contaminant transport model was developed for use in the evaluation of natural attenuation as the preferred remedial alternative at the site. Previous modeling efforts between 1993 and 1997, in combination with a comprehensive field program, had found that pump-and-treat would remove only a small percentage of the total VOC mass and that natural attenuation would be more effective at this site. Previous contaminant transport modeling using MT3D had indicated the potential for natural attenuation to be effective, however biodegradation processes were not adequately simulated for parent and daughter products in a reactive chain, therefore contaminant transport modeling using the biologically reactive, multi-species transport code RT3D (Clement, 1998) was recommended. With the biologically-reactive, multi-species transport code, predictions of the fate of the chlorinated solvents can be better represented due to explicit simulation of the

---

[1]Currently with Marshall Macklin Monaghan Limited, Thornhill, ON, Canada
[2]Currently with Franz Environmental Inc., Bolton, ON, Canada

kinetics of biodegradation in the anaerobic and aerobic zones.

The objective of the reactive transport modeling was to re-evaluate the effectiveness of natural restoration processes in attenuating the chlorinated VOC plumes by implementing new data and to provide input to the evaluation of the need for additional groundwater monitoring or active source remediation.

GROUNDWATER FLOW MODEL

Three-dimensional groundwater flow was simulated using Visual MODFLOW (WHI, 1998). A 4-layer system was simulated where the top 3 layers represent the shallow, intermediate, and deep zones screened by the observation wells. The bottom model layer extends from the deep zone down to a silt layer that was identified during a previous investigation (Dames & Moore, 1994). The simulated aquifer thickness is 125 ft. The sand aquifer is reasonably homogeneous over the scale of interest, with a hydraulic conductivity of 75 ft/day.

A single layer model was also created, as it was determined that the computer time required for reactive transport simulations using a 4-layer model was excessive. The single layer model was created from the 4-layer model. The geometry of the 1-layer model is identical to the 4-layer model in all respects except for the number of layers. The most important difference between the two models is that the rivers and streams are assigned in the 1-layer model throughout the full thickness of the model, and therefore will exert a greater influence on the intermediate and deep model layers than in reality or in a multi-layered model.

A total of 167 monitoring wells were used to calibrate the groundwater flow model. Approximately half of these wells (83) are located in Areas I and J. Both the 4-layer and 1-layer models simulate the observed hydraulic heads well. The overall calibration error for the groundwater flow model is approximately 3%.

REACTIVE TRANSPORT MODEL

The reactive transport model calibration focused on the contaminants PCE, TCE, c-DCE and VC. Other chlorinated solvents (TCA, DCA, 1-1-DCE) have been detected in the groundwater at Areas I and J at NAES Lakehurst, but were not simulated in the reactive transport model. The TCA and DCA distributions in the south plume are very similar to the TCE and c-DCE distributions. The biodegradation rates for TCA and DCA are also similar to those for TCE and c-DCE. Therefore, predictions of natural attenuation for these two groups of contaminants are expected to be similar.

Conceptual Model: The conceptual model for source release at Areas I and J reflects the behavior of DNAPLs in the subsurface. The dissolved groundwater contamination at Areas I and J is greatest in the deeper zone of the aquifer (50-70 ft bgs) within the areas of potential historical contaminant releases. This suggests that DNAPL has migrated vertically to at least this depth. Due to the low concentrations of PCE, TCE, and TCA it is likely that the contaminants currently are present at residual saturation. Based on the relatively minor concentration decreases over the monitoring period (1992-1998), it is likely that dissolution from residual sources at depth is still occurring in some areas of the plumes. Low concentrations of contaminants will continue to be released into the groundwater until the sources have been depleted.

Due to a lack of data regarding potential sources, and a lack of historical monitoring data, assumptions were made with respect to source location, strength and timing. These assumptions define the conceptual model. The conceptual model was then implemented in the numerical model and refined until an acceptable model calibration was achieved.

The development of the conceptual model describing the plumes at Areas I and J at NAES Lakehurst is based in large part on observed contaminant distributions in 1992 and 1998, the earliest and latest dates for which data was available. The PCE, TCE, c-DCE plumes interpreted from wells in the deep aquifer zone are plotted for March 1998 on Figures 1a through 1c.

For the north plume, the source area is assumed to dissolve both PCE and TCE into groundwater from residual DNAPL in the intermediate and deep aquifer zones. The plume, which extends to a downgradient creek, appears to have remained at a fairly steady state since 1992.

The south plume in fact consists of two plumes. The sources that created the south plume are assumed to have been depleted prior to the start of the monitoring program in 1992. Modeled contaminant sources in the south plume are assumed to have consisted of TCE only.

Model Calibration: The single layer model was developed and calibrated to observed contaminant plumes in the deep zone of the aquifer where the highest contaminant concentrations were observed. The 4-layer model was then implemented using the input parameters determined using the 1-layer model. The 4-layer model is important with respect to the assessment of potential plume migration beneath the downgradient creek. Some further calibration was required as, with greater vertical discretization, some contaminant sources and their associated anaerobic degradation zones were located in the intermediate zone and some in the deep aquifer zone.

The model simulates the historical development of the VOC plumes starting in 1958, which was determined to be the earliest date that activities occurred which had the potential to contaminate the groundwater.

In the model, residual sources were located near areas previously identified as potential source areas (Dames & Moore, 1993), however the final location of source areas in the model was determined through model calibration. Source concentrations and the time during which sources remained active were determined through model calibration in accordance with the available information regarding past practices at the site. Sources were implemented using the specified concentration boundary, which provides solute mass to the simulated domain.

Equilibrium-controlled adsorption of contaminants to soil particles, aerobic degradation and anaerobic sequential biological degradation of PCE through TCE, c-DCE, VC, and ethene were simulated. Biodegradation was assumed to occur only in the aqueous phase and negligible degradation of the sorbed phase was assumed.

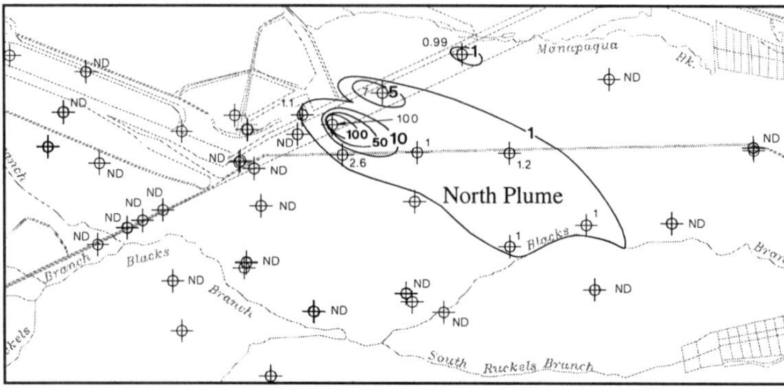

FIGURE 1a - Observed PCE Concentration March 1998

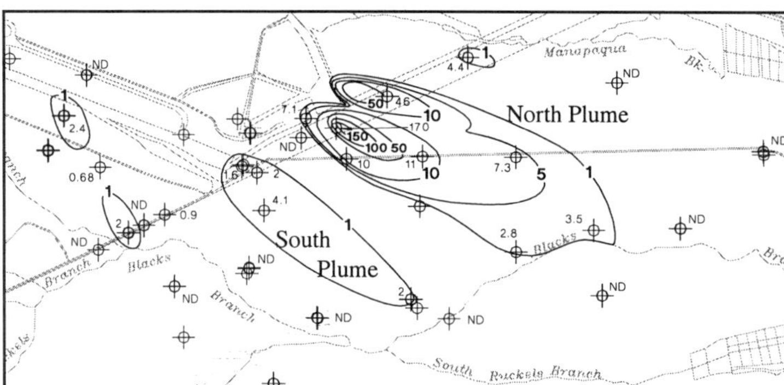

FIGURE 1b - Observed TCE Concentration March 1998

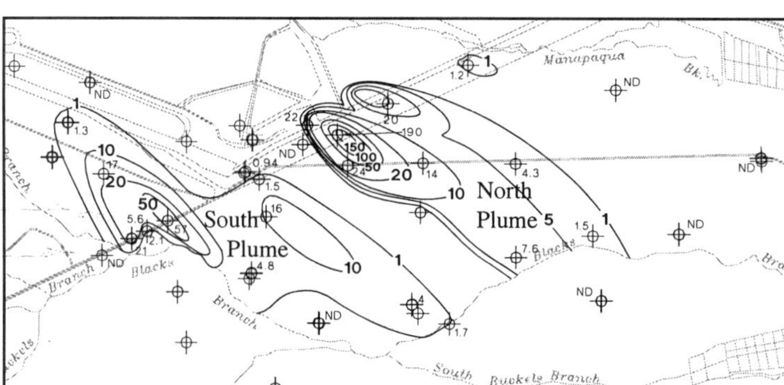

FIGURE 1c - Observed c-DCE Concentration March 1998

Biodegradation rates for PCE, TCE, c-DCE and VC in the anaerobic and aerobic zones calculated from observed concentration data using transect analysis (Table 1) were used as input to the transport model. Source locations, concentrations and contaminant degradation rates were then varied to achieve the best possible numerical solution based on the available data and current conceptual model.

The northern plume was simulated quite well using the biodegradation rates from the transect analysis. However, observed distributions of TCE and c-DCE in the southern plume were difficult to reproduce. Currently, TCE and TCA are almost absent in the groundwater, and have been replaced by their degradation products c-DCE and 1,1-DCA. The model simulates the historical development of contaminant plumes from assumed source locations based on the conceptual model for the site however there is a high degree of uncertainty surrounding the location and history of the sources in the southern plume. Contaminant degradation rates were adjusted in order to calibrate the south plume. Calibrated degradation rates are listed in Table 1 for the north and south plumes. The extent of anaerobic degradation zones in the aquifer are shown on Figures 2a through 2c.

The results of the transport model calibration using the 1-layer model representing the deep aquifer zone shown for March 1998 in Figures 2a through 2c for PCE, TCE and c-DCE can be compared to the observed concentrations for the same date shown in Figures 1a through 1c.

Overall, observed contaminant plumes are well represented in the model. Concentrations at individual wells may be slightly higher or lower than observed concentrations, however the overall shape and concentrations in the north plume are simulated well. For the south plume, the extent of both simulated plumes is greater than the extent of the observed plumes. Therefore, predictions made using the model will be conservative with respect to the extent of the south plume.

The plumes simulated using the 4-layer model are very similar although slightly thinner than the plumes generated using the 1-layer model. Contaminant concentrations are lower through the central portion of the plumes by up to 5 ppb for all species in the 3-D simulation. Vertical dispersion between adjacent layers in the 4-layer model leads to slightly lower contaminant concentrations for the same source loading condition. The 3-D simulation shows that both contaminant plumes cross over to the south side of the downgradient creek.

Table 1. Half-lives for contaminant biodegradation (yr^{-1})

	Calculated	**Modeled - South**	**Modeled - North**
Anaerobic – PCE	1.5	-	1.5
Anaerobic – TCE	1.5	0.5-1.0	1.5
Anaerobic - c-DCE	1.5	-	1.5
Anaerobic - VC	0.1	0.1	0.1
Aerobic - PCE	231	-	231
Aerobic - TCE	21	21	21
Aerobic - c-DCE	5.7	11.4	5.7
Aerobic - VC	0.1	0.1	0.1

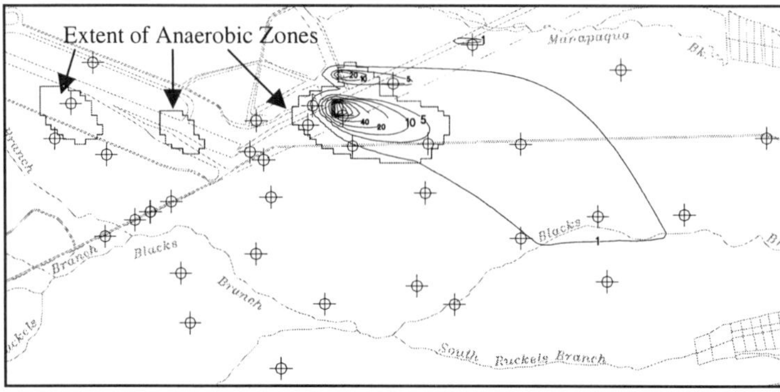

FIGURE 2a - Simulated PCE Concentration March 1998

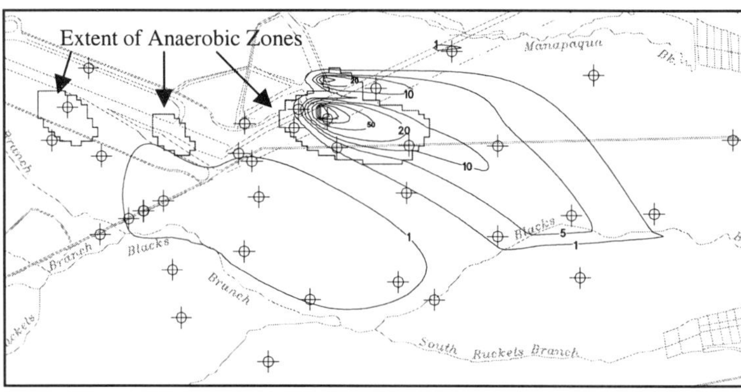

FIGURE 2b - Simulated TCE Concentration March 1998

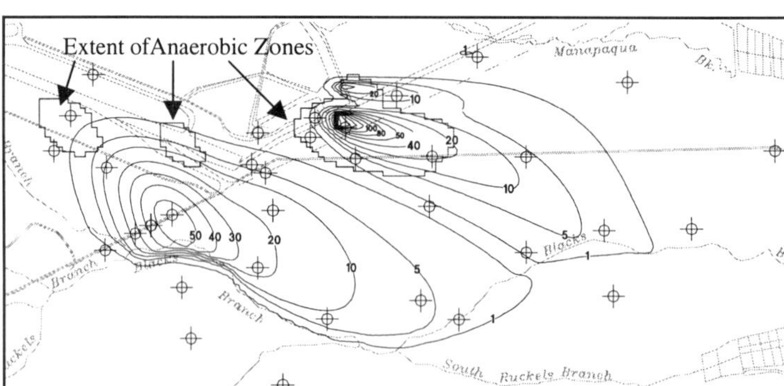

FIGURE 2c - Simulated c-DCE Concentration March 1998

PREDICTIVE SIMULATIONS

The calibrated model was used to predict future migration of the chlorinated VOC plumes. Source concentrations were decreased assuming a 15 year half-life, as was used to calibrate the model. Active sources remain only in the north plume.

Three-dimensional model simulations showed that the extent of the north plume may increase marginally until the sources are fully depleted and that the leading edge of the northern PCE and TCE plumes will cross over to the south side of the downgradient creek at concentrations below 5 ppb, before steadily decreasing due to natural attenuation processes. The model predicts that all contaminant concentrations will be below 5 ppb 15 years after source depletion, and that all contaminant concentrations will be below 1 ppb 25 years after source depletion.

Since the time frame required for natural attenuation to reduce contaminant concentrations to acceptable levels is dependent on the model input parameters, a sensitivity analysis was conducted to address the inherent variability in this time estimate.

MODEL SENSITIVITY

The purpose of the sensitivity analysis was to quantify the uncertainty in the calibrated model caused by uncertainty in the estimates of input parameters. For the reactive transport model, uncertainties are associated mainly with the specified degradation rates. Therefore the sensitivity analysis focussed on the effect of changes in contaminant degradation rates and the source depletion rate on model predictions.

Degradation rates for all species were varied concurrently by a factor of 2 upward and downward to test the sensitivity of the model (plume concentration and life) to variations in the contaminant biodegradation rates applied in the model within the uncertainty of the rate calculation.

The sensitivity simulations were run from 1992 through to 2062. This starting date was chosen to provide for a comparison of the observed and simulated plumes in 1998 which allows some assessment of whether or not each sensitivity run represents a reasonable and possible scenario. If the simulated contaminant distributions in 1998 do not appear reasonable, then the applied degradation rates may not represent a probable scenario.

By varying the biodegradation rates of all species in the north plume, it was shown that a change in the biodegradation rate by a factor of 2 will change the time frame required for natural attenuation by approximately 10 years. It was also determined that the error in the estimated biodegradation rates is likely much less than a factor of 2 as a two-fold increase did not provide a good representation of observed concentrations in 1998.

The sensitivity analysis also showed that natural attenuation will restore the aquifer more quickly if the modeled sources decrease at a faster rate. Using a half-life for source depletion of 5 years instead of 15 years decreased the predicted time required to restore the aquifer by approximately 10 years. Since extrapolation and many somewhat uncertain variables are involved in the simulation of contaminant migration, it is deemed appropriate to rely on the more conservative predictions.

CONCLUSIONS

The single layer reactive transport model calibrated well to observed contaminant plumes in the deep aquifer zone in 1992 and 1998 at Areas I and J at NAES Lakehurst. The model calibrated better when the simulated kinetics of biodegradation were different for the north and south plumes, possibly due to uncertainties associated with the contaminant sources and degradation rates for the south plume.

The reactive transport parameters determined through calibration of the single layer model were implemented in a three-dimensional model simulation. A comparison of the calibration in the deep aquifer zone showed that a similar calibration was achieved as in the single layer model. The same time frame required for aquifer restoration was predicted by both of the models.

Based on the current understanding of contaminant migration and degradation in Areas I & J, the calibrated model can be used to realistically predict the transport of PCE, TCE, c-DCE, and vinyl chloride in the groundwater. The model was used to predict the future extent of plume migration, assuming continued dissolution of contaminant mass into the groundwater from residual sources. Model simulations showed that the contaminant concentrations in the north plume will decrease steadily due to natural attenuation processes, although the extent of the contaminant plume may increase marginally until contaminant sources are fully depleted.

A sensitivity analysis provided some measure of the uncertainties associated with the reactive transport modeling. A change in the biodegradation rate by a factor of 2 was found to change the time frame required for natural attenuation by approximately 10 years. The sensitivity analysis also showed that natural attenuation will restore the aquifer more quickly if the modeled sources decrease at a faster rate. Although active source remediation would reduce the time frame for restoration, residual sources are at depth and are not well defined.

Given sufficient time, natural restoration of the aquifer will occur and monitored natural attenuation remains a highly viable remedial strategy for the chlorinated VOC plumes at this site.

REFERENCES

Clement, T.P., 1998. RT3D Manual - *A Modular Computer Model for Simulating Biologically-Reactive Multi-Species Transport in Three-Dimensional Saturated Groundwater Systems*, PNNL, April 1998.

Dames & Moore Inc., 1993. *Aquifer Characterization - Areas I & J, Volume 1, Naval Air Warfare Center - Aircraft Division, Lakehurst, NJ*, February 15, 1993.

Dames & Moore Inc., 1994. *Pre-design Groundwater Remedial Investigation, Areas I & J, Naval Air Engineering Station, Lakehurst NJ*, April 1, 1994.

WHI, 1998. *Visual MODFLOW User's Manual*, Waterloo Hydrogeologic Inc.

INTRINSIC BIODEGRADATION OF VINYL CHLORIDE IN THE VADOSE ZONE

Duane Graves, Ph.D. and Dirk Pohlmann, IT Corporation, Knoxville, TN
Sherry Ingram, IT Corporation, Lake Charles, LA, USA

ABSTRACT Natural attenuation has become a critical component of many groundwater management strategies and remedial actions. The processes and methodology for evaluating natural attenuation in groundwater has been established by the US Air Force, the US EPA, and several state regulatory agencies. The application of natural attenuation to the vadose zone is not as well documented and often relies on a significantly different approach to document the processes acting to reduce contaminant mass and concentration. Historical and recent soil gas data were used to define the contribution that natural attenuation was having on vadose zone restoration. The area and volume of a vinyl chloride impacted vadose zone was determined geostatistically using data collected at four time points spanning seven years. Changes in the relative mass of vinyl chloride in the vadose zone and the change in the total area impacted by vinyl chloride were used to demonstrate soil restoration by natural attenuation. The mass of vinyl chloride had decreased by 96 percent during the past seven years. The results also indicated a 2-fold expansion in the vinyl chloride plume during the same time. The rate of mass reduction observed during the past six years indicated that the vinyl chloride plume was attenuating with a half-life of 1.2 to 1.8 years. Soil gas composition demonstrated that biodegradation was largely responsible for the change in mass. Evidence for aerobic and anaerobic biodegradation of vinyl chloride was observed in the soil gas. The oxygen concentration was depressed, the carbon dioxide, ethene, and ethane concentrations were elevated, and methane was being produced within the impacted area. The results from this evaluation provided strong evidence that natural attenuation was effectively remediating the impacted vadose zone. The evaluation approach seems reliable for vadose zone plumes that have multiple sampling events containing enough sample locations to delineate plume dimensions and the changes that occur over time.

INTRODUCTION

A pipeline leak resulted in a vinyl chloride (VC) release into the vadose zone in 1992. The vadose zone at the site is between 2 and 5 feet thick and is bounded by the water table on the bottom and a less-permeable limey clay layer near the surface. Because VC has a large Henry's Law constant, it was predicted to strongly partition to the gas phase and not dissolve to high concentrations in water. The geologic setting and physical properties of VC minimize potential migration pathways, such as releases to the atmosphere or groundwater, and make horizontal diffusion the primary mechanism for movement of VC.

A field monitoring program is underway to evaluate the natural attenuation of vinyl chloride (VC) in the vadose zone. Following the detection of the VC release along the VC pipeline in July of 1992, several investigations were conducted to delineate the distribution of chlorinated hydrocarbons. The pipeline was taken out of service in August 1992. Data collected from 44 locations in October 1992 yielded the maximum VC concentrations detected to date and are considered to represent the maximum amount of VC in the vadose zone. In 1997, 31 temporary soil gas sample points were placed throughout the VC-impacted area to assess the change in the VC plume. In July 1998, soil vapor samples were collected from the same 31 locations sampled in 1997. New temporary vapor monitoring points (29 points) were installed in 1999. The soil vapor data from each sampling event were compared to evaluate the attenuation of VC. The vadose zone evaluation for the site employed relative-mass and plume-area calculations, soil gas composition, and statistical correlation tests to conclude that naturally-occurring biodegradation of VC was making a clear and appreciable contribution to the overall attenuation of VC in the area. Groundwater was collected in 1999 from two sampling points to examine the partitioning of VC into the aquifer.

METHODS

Documenting natural attenuation in the vadose zone requires a slightly different approach than that used for demonstrating natural attenuation in groundwater. Many of the parameters measured in groundwater cannot be reliably measured in soil. Examples include nitrate, sulfate, chloride, and ferrous iron. These parameters are strongly associated with water and leach out of the soil without leaving a useful signature for measuring natural attenuation. A reliable method for measuring redox in soil in not available. Thus many of the typical natural attenuation parameters do not provide useful information in vadose zone soil.

Ethene, ethane, methane, CO_2, and O_2 are useful in establishing the effectiveness and extent of naturally-occurring biodegradation of VC in the impacted area. Soil gas samples were collected from the impacted vadose zone in October 1992, June 1997, July 1998, and June 1999. Soil gas samples were analyzed for VC in 1992, for VC, ethene, ethane, and methane in 1997, and for VC, ethene, ethane, methane, O_2, and CO_2 in 1998 and 1999. Results were used to generate two-dimensional VC plume maps for 1992, 1997, 1998, and 1999. An exponential kriging algorithm (Keckler, 1995) was used to generate the VC contours and to calculate plume volume and area with concentrations above 10 microliters per liter (μL/L) for each data set. The soil gas data were also examined for statistically significant correlations using the Pearson Product-Moment Correlation Coefficient method (Statistica for Windows, StatSoft, Inc., 1998).

RESULTS AND DISCUSSION

The soil gas VC data for 1992, 1997, 1998, and 1999 were examined for VC mass, areal extent, and correlation with degradation products. The 1999 soil gas data (Table 1) were consistent with expectations for soil gas influenced by intrinsic biological activity. Examination of the soil gas data and correlations between various gases provide convincing evidence that biological activity is influencing the composition of soil gas in the VC-impacted area. Soil gas samples from 1997, 1998, and 1999 were analyzed for methane, ethene, ethane, O_2, and CO_2 (Table 1).

Ethene and ethane are potential byproducts of the reductive dechlorination of VC. Under anaerobic conditions, their presence strongly indicates VC dechlorination. Correlation analysis of ethene and ethane with the presence of VC have indicated a statistically significant correlation, suggesting a relationship between the dechlorination of VC and the production of ethene and ethane. Methane was routinely detected in area soil and fugitive emissions from a natural gas mixing station within the impacted area may be contributing to methane detections at some of the sample locations. Methane at the site does not appear to be directly related to microbial processes associated with the VC impacted area. Soil gas O_2 was generally depleted in areas with high concentrations of VC, methane, and CO_2. However, the data were variable and a significant relationship did not appear to exist between O_2 and VC, ethene, or ethane. Carbon dioxide was elevated in several locations. High concentrations of VC always coincided with high carbon dioxide concentrations; however, high CO_2 concentrations were also observed in some locations with low concentrations of VC. The observation of elevated CO_2 concentrations within the VC plume suggests that VC is being completely biodegraded to CO_2, water, and chloride ions; however, the elevated concentration of CO_2 in uncontaminated areas indicates a generally high level of indigenous microbial respiration.

TABLE 1. Soil Gas Results

Sample Location	VC (µL/L)	Methane (µL/L)	Ethane (µL/L)	Ethene (µL/L)	O2 (percent)	CO2 (percent)
1	0.8	179	0.5	0.1	3.5	ND
2	1.0	116141	149.9	14	9.0	2.1
3	ND	6	0.7	0.3	6.9	10.1
4	0.5	180314	3.0	0.1	2.6	8.6
5	6.0	74944	24	2.3	3.3	17.7
6	0.6	160243	23	1.7	3.6	8.4
7	0.5	18585	0.5	0.1	17.3	ND
8	ND	604400	20	8.5	3.7	ND
9	0.6	63	0.4	0.1	10.3	0.2
10	0.8	123	2.7	1.7	14.5	1.3
11	ND	8858	81	9027	8.8	6.7
12	ND	11039	70	3	5.9	10.3
13	ND	7297	382	2	14.5	1.5
14	3	37154	1851	103	5.7	11.7
15	0.5	6	0.7	0.5	10.2	7.1
16	ND	29	0.2	0.1	5.2	14.1
17	77	175585	6481	230088	18.2	22.0
18	0.6	927	26	81	8.4	11.8
19	1647	1828	672	2511	4.1	10.5
20	0.6	4400	122	3	15.8	4.9
21	0.5	229	12	38	17.0	0.3
22	30.2	608	416	22853	9.9	10.7
23	3281	7939	1980	12462	3.4	13.8
24	0.5	330	0.1	0.1	8.3	0.9
25	0.7	1394	12	2	5.2	6.1
26	945	183404	4120	1348	2.0	11.7
27	0.6	9589	0.6	2	13.9	1.8
28	0.5	2	0.2	0.1	11.8	ND
29	1.3	163685	8.6	2	5.3	13.3
10 (blank)	ND	2	0.1	0	16.8	ND
20 (blank)	ND	2	0.1	0.3	17.5	ND
29 (blank)	ND	2	0	0	17.5	ND

ND, not detected

Examination of the soil gas data indicates that VC concentrations have steadily declined since October 1992. Plume dimensions contoured using 1992, 1997, and 1998 data were compared to the plume dimensions from the 1999 data (Figure 1). As sampling events from 1992 had established plume closure at the building to the north, the plume contours generated for 1999 were not defined to the north, and the kriged plume contours have been terminated (Figure 1). In the absence of soil gas samples to the north, the expression of the kriged plume to the north is exaggerated, resulting in an over-estimation of impacted area and relative VC mass for 1999. Relative proportional values were used to quantitatively evaluate changes in VC mass and areal extent since 1992. Both mass and areal extent were set to 100 percent for the 1992 results. Results from 1997, 1998, and 1999 were normalized to the 1992 data. The relative mass of the plume has declined dramatically since 1992 with no more than 4 percent of the original mass still remaining in the vadose zone (Figure 2, mass shown by circles). The areal extent of the plume seems to have reached a maximum and has begun to recede. Relative changes in mass and areal extent of the VC plume were used to approximate the future behavior of the plume (Figure

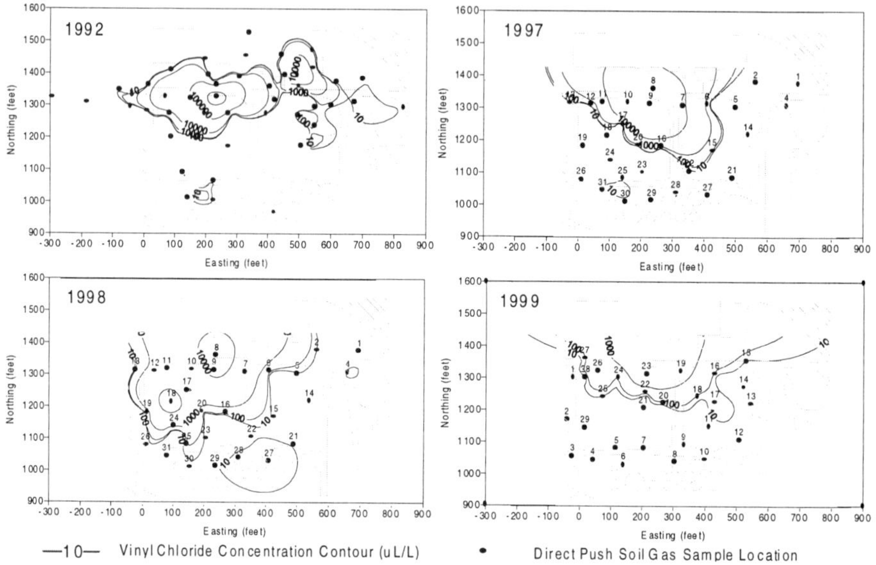

FIGURE 1. Distribution of Vinyl Chloride in the Vadose Zone

2). Regression analysis of the decay of VC mass yielded an apparent attenuation rate of 0.378 per year, resulting in a half-life of 1.8 years.

The decay of VC mass is expected to continue to approximate the dashed line shown in Figure 2. Further refinements of the attenuation rate estimate will be made as additional data become available. The areal extent of the VC impacted soil has increased by slightly more than two-fold from 1992 to 1998, and the plume appeared to be expanding. However, based on 1999 results, this trend seems to be reversed and the areal extent of the plume appears to be decreasing (Figure 2, plume area shown by squares).

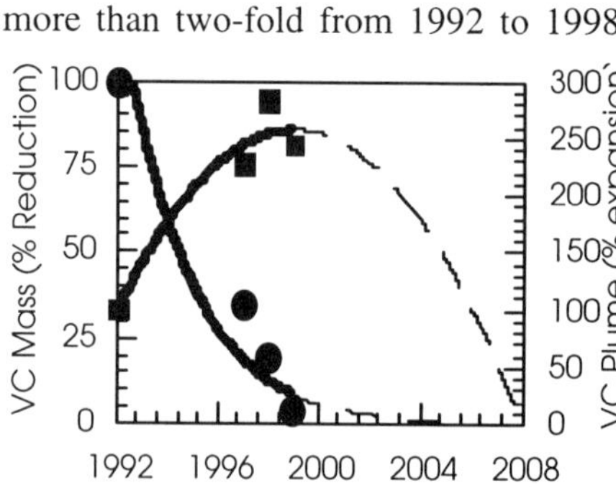

FIGURE 2. VC plume dimensions, relative mass, and areal extent

A simple rank ordering of concentration data (Table 2) documents the rate of concentration change occurring in the soil. The soil gas data for each year was ordered by rank from the highest concentration to the lowest concentrations. The first 22 highest ranking samples were then examined for trend and the rate of change was calculated using an exponential regression. The results consistently show that concentrations have declined every year. The rate of VC concentration decrease ranges from -0.37 per year to -1.08 per year. The average rate of concentration change was -0.64 per year with an average half-life of 1.2 years which compares favorably with the rate calculated by regression analysis of mass reduction.

TABLE 2. Rank Order Comparison of Soil Gas Vinyl Chloride Concentrations, 1992-1999

Rank Order	Oct-92 (μL/L)	Jun-97 (μL/L)	Jul-98 (μL/L)	Jun-99 (μL/L)	Concentration Change Rate (per year)	Concentration Half-Life (years)
1	936000	80123	20579	3,281	-0.77	0.9
2	101736	42067	16753	1,647	-0.49	1.4
3	84672	19433	13779	945	-0.53	1.3
4	63360	11202	5524	76.67	-0.77	0.9
5	51480	3529	2889	30.24	-0.88	0.8
6	47088	52.4	2569	5.99	-1.08	0.6
7	3960	43.0	1751	2.69	-0.78	0.9
8	1580	23.5	171	1.26	-0.84	0.8
9	1264	22.9	106	0.97	-0.85	0.8
10	954	21.6	104	0.81	-0.82	0.8
11	526	17.2	84.7	0.76	-0.75	0.9
12	454	7.90	27.2	0.74	-0.80	0.9
13	100	5.10	21.5	0.62	-0.59	1.2
14	81	3.90	20.4	0.59	-0.57	1.2
15	68	3.80	14.4	0.58	-0.56	1.2
16	36	3.00	10.4	0.58	-0.48	1.4
17	32	2.80	10.4	0.58	-0.46	1.5
18	31	2.30	9.9	0.51	-0.48	1.5
19	17	1.72	8.8	0.49	-0.39	1.8
20	16	1.45	6.5	0.48	-0.41	1.7
21	12	1.40	5.2	0.48	-0.37	1.9
22	11	1.20	4.1	0.479	-0.37	1.9
				Average	-0.64	1.2

The highest VC concentration observed in 1999 (3281 μL/L) represents a 99.7 percent reduction compared to the highest VC concentration in October 1992 (936,000 μL/L), and a continuing decrease from maximum concentrations in 1997 (80,100 μL/L) and 1998 (20,600 μL/L). In 1992, sample points located close to the VC leak had concentrations ranging between 51,000 and 102,000 μL/L. By the 1999 sampling event, no sample point contained a VC concentration in this range, and maximum concentrations were all less than 5000 μL/L. The soil gas composition, change in mass and areal extent, and Pearson Product-Moment Correlation analysis consistently support the conclusion that VC is attenuating and that intrinsic biodegradation is a significant component of the attenuation process.

Two groundwater samples were collected in 1999 to examine the dissolution of VC from the soil gas into groundwater. The partitioning of VC from soil gas into groundwater is expected to be limited due to the large Henry's Law constant for VC (3530 x 10^2 atmosphere cubic-meters per mole compared with 10 x 10^2 atmosphere cubic-meters per mole for trichloroethene) (Chidgopkar, 1996). In an undisturbed situation, VC is predicted to equilibrate between vapor and liquid at a concentration proportional to its Henry's law constant; that is, the Henry's law constant times the dissolved concentration in moles is predicted to equal the concentration of VC in vapor in atmospheres:

$$P_a = H_aC_a$$

where: P_a = the concentration of VC in soil gas (atm);
H_a = the Henry's law constant for VC (81.9 atm/M); and
C_a = concentration of VC in groundwater (M).

Groundwater samples were collected from vapor monitoring points 20 and 23 and analyzed for dissolved VC using US EPA Method 8240. By solving this equation for C_a and converting the result to mg/L, a direct evaluation can be made concerning the equilibration of VC between the soil gas and the groundwater. The results, based on soil gas concentrations of 0.58 µL/L at location 20 and 3281 µL/L at location 23, predicted dissolved VC concentrations of 0.00044 and 2.504 mg/L, respectively, at a groundwater temperature of 25°C (Table 3). The actual measured concentrations of VC in the groundwater at sample locations 20 and 23 were 0.037 and 2.4 mg/L respectively. The extremely close correspondence between the concentration measured as location 23 and the predicted groundwater concentration suggests that VC is in equilibrium between the soil gas and the groundwater.

The calculation performed on data from sample location 20 yielded an expected groundwater VC concentration much lower than the actual measured concentration, suggesting that the soil gas and groundwater concentrations of VC are not in equilibrium at this location. Elevated values may reflect increased concentrations due to previously higher soil gas values. Also, the difference between the calculated and measured values may not be as extreme as it appears as the measurement of low concentration of VC in both soil gas and groundwater is more sensitive to analytical errors than the measurement of high concentrations.

TABLE 3. Predicted and Observed Groundwater VC Concentration

Sample Location	Predicted C (mg/L)	Measured C (mg/L)
Point 23	2.504	2.4
Point 20	0.00044	0.037

The close correspondence between the predicted and measured groundwater concentrations at location 23 suggests that VC is in equilibrium between the soil gas and the groundwater nearer to the release. The equilibrium relationship between vapor and dissolved concentrations of VC at sample location 23 provides support for the conclusion that remediation of one phase will naturally lead to the remediation of the other phase. Where natural attenuation processes are effectively removing VC from the vadose zone, restoration of the vadose zone should coincide with the restoration of shallow groundwater. The close agreement of the predicted and measured VC concentration in groundwater at location 23 provides strong support for this conclusion.

Natural attenuation of the VC plume in the vadose zone is being brought about by biological and physical (dispersive) processes. These results strongly support the conclusion that naturally occurring biodegradation is a significant component of the overall attenuation of VC in the vadose zone at the area. Although the VC plume has dispersed to an extent, the mass is being dramatically reduced by destructive biological processes and the plume remains on-site. The geologic setting in the area restricts the vertical movement of VC and minimizes pathways to potential receptors. The biodegradation processes currently operating are expected to be self-sustaining as long as VC is present. Therefore, natural attenuation continues to be an extremely viable remedial technique for this VC-impacted vadose zone.

REFERENCES

Chidgopkar, V.R., 1996, "Applying Henry's Law to Groundwater Treatment," Pollution Engineering Online, http://www.manufacturing.net/magazine/polleng/archives/1996/pol0301.96/03adp1f0.htm.

Keckler, D., 1995, "Surfer For Windows Version 6," Golden Software, Inc., Golden, CO.

StatSoft, Inc., 1998, STATISTICA for Windows, Tulsa, OK.

NATURAL ATTENUATION ASSESSMENT OF A HIGHLY IMPACTED TRICHLOROETHENE SITE

by ***Jack. T. Spadaro***, Charles T. Esler, Paul M. Trone, and Michelle L. Peterson
(AGRA Earth & Environmental, Inc., Portland, Oregon)

ABSTRACT: Chlorinated ethenes (CEs) have been used extensively in industry and science. The recalcitrance of the compounds to degradation, their carcinogenic potential, and poor historic disposal practices have led to heightened regulatory scrutiny and public awareness. A chlorinated hydrocarbon plume extends from approximately 20 feet (6 m) below ground surface (bgs) to approximately 60 feet (23 m) bgs, and approximately 180 feet (55 m) laterally from the release point at the subject site (Site) in Portland, Oregon. The primary CEs are trichloroethene (TCE), *cis*-1,2-dichloroethene (*c*-1,2-DCE), and vinyl chloride (VC). A remedial action is planned for the saturated zone. Natural attenuation processes have been monitored in order to prepare for post-remedial management, as well as to potentially estimate the age, volume, fate, and transport of the release(s). The field and laboratory data presented below indicate that significant reductive transformation of TCE to daughter products has occurred in the Site subsurface.

INTRODUCTION AND BACKGROUND

The currently vacant 3.34-acre Site is the former location of a medical laboratory that operated from 1961 to 1979. Medical laboratory operations, as well as subsequent suspected illegal waste disposal practices, may have released TCE to the subsurface in the vicinity of the Site's dry wells. Subsurface exploration from the late 1980s through 1999 revealed the presence of TCE, *c*-1,2-DCE, and VC concentrations as high as 550 milligrams per liter (mg/L), 370 mg/L , and 27 mg/L, respectively, in the Site groundwater.

The Site is located in the Columbia River floodplain. Site geology from the near surface to approximately 60 feet (18.5 m) bgs consists of fluvial and lacustrine depositional sequences (Overbank deposits). Silts and sands are discontinuously interlayered throughout, and often contain organic matter. Groundwater is encountered in silts approximately 10 feet (3 m) bgs. Sands, generally encountered below 35 feet (10.7 m) bgs, are fine- to very fine-grained, well-sorted, and contain volcanic cinders. Underlying the Overbank deposits is the Troutdale Gravel Aquifer (TGA) unit, between 60 and 235 feet (18 and 72 m) bgs, which consists of unconsolidated and cemented gravels.

Natural attenuation (intrinsic remediation) has become increasingly recognized as a remedial tool for treating CEs dissolved in groundwater. Natural attenuation processes have been defined by the Environmental Protection Agency to include biodegradation, dispersion, dilution, sorption, volatilization, and chemical and biochemical stabilization of contaminants. These processes can lead to losses in contaminant toxicity, mobility, or volume.

Biodegradation is often recognized as the dominant destructive attenuation mechanism. TCE biodegradation occurs most readily under anaerobic electrochemically reducing conditions by reductive dechlorination (RD) processes, (e.g. electron and hydrogen addition to a TCE molecule, with displacement of a chlorine atom as chloride). The TCE is primarily transformed to *c*-1,2-DCE, which can be further sequentially transformed to VC and ethene. The transformation of *c*-1,2-DCE to VC and ethene is relatively difficult under anaerobic conditions, and *c*-1,2-DCE and VC can accumulate.

The full consideration of natural attenuation as a remediation option requires two or more of the following lines of evidence, with the first absolutely required: 1) reduction in contaminant concentrations along the flow path downgradient from the contaminant source; 2) loss of contaminant mass at the field scale, using chemical and geochemical data, conservative tracer studies and modeling to demonstrate mass reduction and in-situ attenuation rates; and, 3) microbiological laboratory data that support the occurrence of biodegradation and yield biodegradation rates. The Site monitoring data and the laboratory studies summarized here fulfill the first line of evidence, and partially fulfill the second and third lines of evidence.

MATERIALS AND METHODS

Field Study. Data was collected during September 1998 (dry season) and March 1999 (wet season) monitoring events on fourteen monitoring wells chosen to represent the plume center, downgradient and perimeter areas in the Overbank and TGA zones. Thirteen of the wells are shown in Figure 1. The groundwater dissolved oxygen (DO) concentration, oxidation reduction potential (ORP), pH and temperature were measured in the field during sample collection using micropurge techniques. Groundwater samples were analyzed for presumptive organic and inorganic products of TCE biodegradation, including *c*-1,2-DCE, VC, ethene, and chloride. Samples also were analyzed for various constituents that mediate biological activity and reductive capacity in the impacted media, such as: the aromatic compounds benzene, toluene, ethylbenzene, and xylenes (BTEX); the electron acceptors sulfate and nitrate; total organic carbon (TOC); ferrous (Fe^{2+}) and ferric iron (Fe^{3+}) species; and nutrients. The monitoring parameters chosen were based upon those proposed by the EPA (Wiedemeier, et al., 1998).

Laboratory Biofeasibility Testing. Limited laboratory degradation tests on Site groundwater were performed in late 1997. These included anaerobic microbial counts, biological activity screens, and an anaerobic treatment study.

Most probable number (MPN) counts were performed on groundwater samples collected at monitoring wells MW-29 and MW-31. The MPN counts assessed the populations of anaerobic microorganisms viable in the presence of both a carbon source amendment (either ethanol or acetate) and TCE. Biological Activity Reaction Tests (BARTs) (Hach Co., Loveland, Colorado) were used to qualitatively assess the presence of sulfate-reducing bacteria (SRB) (Hach No. 24324-09) and iron-related bacteria (IRB) (Hach No. 24323-09) in groundwater. Observations made

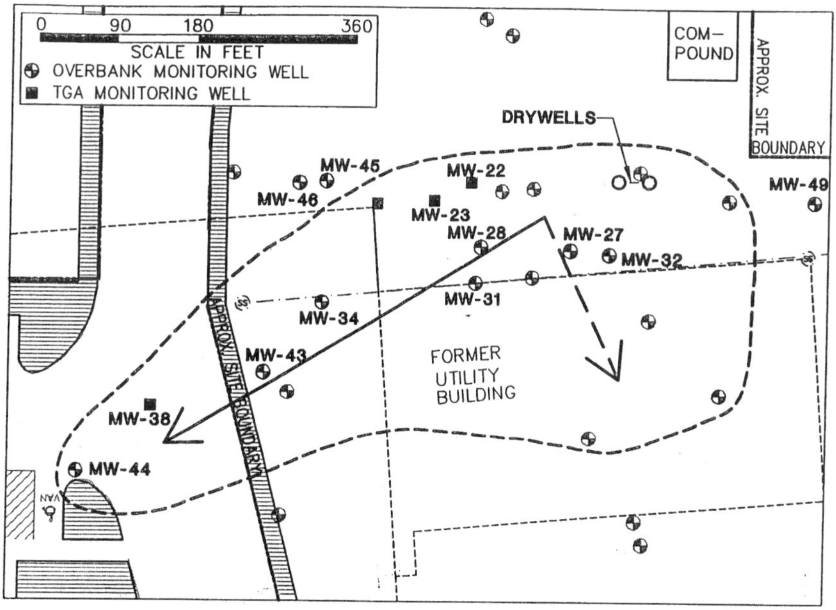

FIGURE 1. Site plan showing placement of monitoring wells, approximate CE plume dimensions (-----), and general groundwater flow directions (--->).

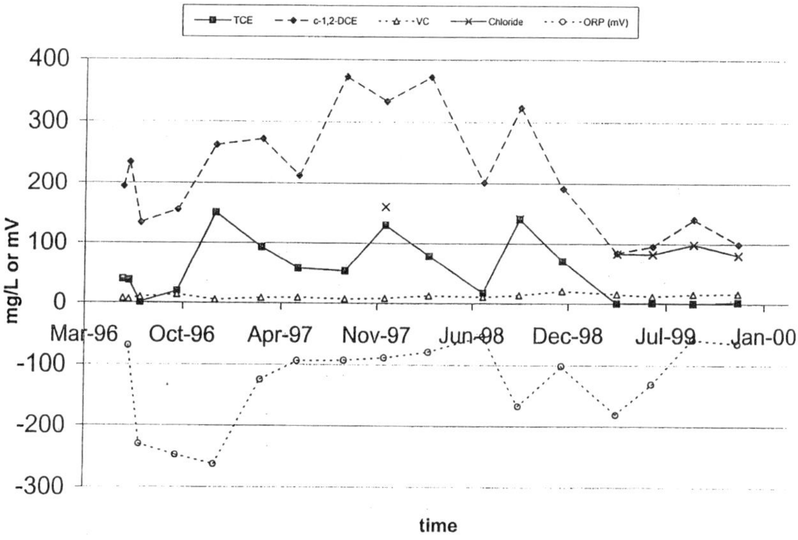

FIGURE 2. Chlorinated ethene and chloride concentrations (mg/L), and ORP (mV) versus time in groundwater at monitoring well MW-27.

of the SRB and IRB BART reactors over a 6-day test period were compared to the standard comparator chart and reaction pattern signature key supplied by the Hach Co.

A degradation study was performed on groundwater collected at monitoring well MW-29. The groundwater was sparged, spiked with TCE to a starting concentration of approximately 6 mg/L, and then incubated under the following conditions: a) native anaerobic; b) killed native anaerobic control; b) ethanol-amended anaerobic; c) ethanol- and sulfate-amended anaerobic; and e) native aerobic. Ethanol was added at a 2 millimolar (mM) concentration, and sodium sulfate was added at a 0.5 mM concentration. Anaerobic batches were mixed and dispensed into 2-milliliter (mL) glass ampules in an anaerobic chamber. Aerobic samples were placed into 2-mL ampules under ambient oxygen conditions. Killed controls were prepared by addition of sodium azide at a 2 mg per mL concentration. The sealed ampules were stored in the dark at approximately 20 °C. Duplicate samples of each of the anaerobic batches were analyzed on Day 0, Day 9, and Day 16. Duplicate samples of the aerobic batch were analyzed on Day 16.

RESULTS AND DISCUSSION

Field Study. There are significant reductions in CE concentrations along the flow path downgradient of the dry wells in the Overbank zone. Results from the TGA zone will not be discussed. The CE concentrations fall by orders of magnitude within lateral distances of 30 to 40 feet, as seen in comparisons of data from samples collected at monitoring wells MW-27 and MW-28, and MW-31 and MW-34. Select monitoring data from selected Overbank wells are summarized in Table 1.

Significant transformation of TCE to daughter products also was indicated, and *c*-1,2-DCE and VC were present in several wells at relatively high concentrations. Accumulation of VC has been suggested for groundwater environments where methane is present at concentrations greater than 0.5 mg/L due to a slower degradation rate for VC than for TCE (Weidemeier, et al. 1998, p 27, p 29). Groundwater samples from monitoring wells MW-27 and MW-31 contained methane at 8.2 and 11.0 mg/L, respectively. VC also apparently underwent conversion to ethene via RD. Ethene concentrations ranged from 1.1 to 3.3 mg/L.

The background chloride concentration for the Site groundwater has been estimated to be in the range of 1 to 20 mg/L. Chloride was present in several groundwater samples at concentrations approximately seven or more times greater than background, a strong indicator that RD has occurred in the Overbank zone of the Site, especially in the vicinity of monitoring wells MW-27 and MW-31.

The geochemical data indicate that the groundwater is strongly reducing in character. The strength of the reducing environment was reflected in: 1) dissolved oxygen concentrations of generally 0.4 mg/L or less; 2) strongly negative ORP readings of -100 mV or lower; 3) methane concentrations of greater than 1.8 mg/L in many instances; and, 4) Fe^{2+} concentrations ranging as high as 32 mg/L. The presence of BTEX at concentrations greater than 0.1 mg/L has been suggested to contribute to the creation of reducing environments (Wiedemeier, et al., 1998).

TABLE 1. Summary of selected analytical results for Overbank groundwater. All results as mg/L, except ORP as mV.

Well	TCE	c-1,2-DCE	VC	CH_4	Ethene	BTEX	TOC	Cl^-	$CaCO_3$	SO_4^{2-}	Fe^{2+}	D.O.	ORP	Fe (tot.)
September 1998														
MW-49	0.001 U	0.001 U	0.001 U	4.7	0.001 U	0.006 U	2.4	1.1	154	13	0.05 U	0.7	-143	19.3
MW-32	0.001 U	0.005	0.003	7.6	0.003	0.006 U	2.3	1.1	121	5 U	3.7	0.4	-144	15.2
MW-27	140	320	14	8.2	2	≥5.6	18.3	140	221	5.9	14	0.6	-167	50.6
MW-28	0.02 U	4.3	1.1	8.9	0.027	0.11 U	3.9	6.4	153	5 U	23	0.1	-164	25.6
MW-31	0.2 U	230	19	11	2.1	≥2.2	18.3	140	210	5 U	2.3	0.4	-165	48.7
MW-34	0.001 U	0.015	1.1	27	2.4	≥0.01	7.9	38	208	5 U	2.2	0.8	-135	60
MW-43	0.001 U	0.001 U	0.001 U	1.8	0.001 U	0.006 U	2.8	2.1	312	5 U	0.05 U	0.2	-176	7.07
MW-44	0.001 U	0.001 U	0.002	9.7	0.042	0.002	29.9	2.1	595	5 U	1.2	0.2	-134	15
MW-45	0.001 U	0.001 U	0.001 U	8.3	0.001 U	0.006 U	2.8	2.9	164	5 U	0.05 U	0.1	-172	19.3
March 1999														
MW-49	0.001 U	0.001 U	0.001 U	6	0.001 U	0.006 U	0.5 U	7.3	160	5	0.2 U	0	-192	14.2
MW-32	0.001 U	0.004	0.002	0.014	0.001 U	0.006 U	1	4.4	115	0.2 U	0.2 U	0.4	-162	11.4
MW-27	0.5 U	84	16	4.4	3.3	≥ 3.2	26.9	83.2	318	0.2	32.7	0	-181	54.4
MW-28	0.1 U	19	2.4	3.4	0.065	≥ 0.09	6.5	21	148	0.4	0.2 U	0	-193	19.1
MW-31	0.2 U	37	12	6.2	1.1	≥ 0.430	7.2	37.8	136	0.2 U	0.2 U	0	-207	17.5
MW-34	0.01 U	3	7.1	20	3	≥ 0.042	12.9	42	178	1	5.4	0	-164	62.9
MW-43	0.001 U	0.001 U	0.001 U	3.7	0.002	0.006 U	1.4	13	312	1.5	5.9	0.2	-184	22.7
MW-44	0.001 U	0.001 U	0.003	2.6	0.007	≥ 0.006	9.8	14.5	210	0.3	1.7	0.2	-172	14.2
MW-45	0.001 U	0.001 U	0.001 U	0.001 U	0.001 U	0.006 U	0.5 U	9.7	150	1.3	0.3	0.3	-172	10.7

Notes:
TCE: Trichloroethene
c-1,2 DCE: cis-1,2-Dichloroethene
VC: Vinyl Chloride
CH_4: Methane
BTEX: Benzene, Toluene, Ethylbenzene, Xylenes
TOC: Total Organic Carbon
Cl^-: Chloride
$CaCO_3$: Calcium carbonate
SO_4^{2-}: Sulfate
Fe^{2+}: Ferrous Iron
D.O.: Dissolved Oxygen
ORP: Oxidation-Reduction Potential
Fe (tot.): Total Iron
mg/L: Milligrams per liter
mV: Millivolts
U/J: Not detected at cited concentration/Estimated value

BTEX compounds were present in groundwater at monitoring wells MW-27 and MW-31 at concentrations ranging as high as 5.6 mg/L.

Wiedemeier et al. (1998) presented a scoring method that weighs the relative contributions of site factors to CE bioattenuation. Incorporating measurements from the most highly impacted portion of a plume, a score in excess of 20 points reportedly provides strong evidence for CE bioattenuation. Based upon the described method, the September 1998 and March 1999 monitoring data support scores of 25 and 29 points, respectively, at monitoring well MW-27, and 26 and 23 points, respectively, at monitoring well MW-31.

The groundwater pH ranged from 6.0 to 7.5 standard units and groundwater temperatures ranged from 9 to 13.5 °C. Groundwater ammonia nitrogen concentrations ranged as high as 6.8 mg/L, and orthophosphate phosphorus concentrations ranged as high as 1.7 mg/L. These nutrient concentrations appeared to be adequate to support biodegradation of the available organic carbon in the regions of the greatest CE impact (monitoring wells MW-27 and MW-31).

Figures 2 and 3 depict historical groundwater monitoring data collected at monitoring wells MW-27 and MW-31, respectively. At MW-27, decreases in TCE concentration appeared to generally coincide with observed decreases in ORP. Initially, the *c*-1,2-DCE concentration in monitoring well MW-27 increased, coinciding with an elevated concentration of chloride. Later, *c*-1,2-DCE concentrations decreased significantly. VC appeared to remain at or near a steady state concentration.

It has been suggested that VC can be degraded under iron-reducing conditions provided there is a sufficient concentration of bioavailable Fe^{3+} (Wiedemeier, et al. 1998). The total iron content of groundwater at the Site is generally much higher than the Fe^{2+} content, suggesting that Fe^{3+} is abundant. The RD and iron-mediated degradation pathways may function to keep VC concentrations at or near steady state.

A significant mass of *c*-1,2-DCE was present in the vicinity of monitoring well MW-31 in 1998 (Figure 3). The *c*-1,2-DCE concentration decreased significantly in late 1998 in tandem with the ORP, and chloride concentrations increased significantly. VC concentrations also showed a slight increase.

Laboratory Biofeasibility Testing. Heterotrophic anaerobes were present at populations of 3.62×10^2 and 3.62×10^3 MPN/mL in groundwater samples from monitoring wells MW-29 and MW-31, respectively. Groundwater samples from the monitoring wells yielded bacterial populations of 2.6×10^1 and 2.84×10^5 MPN/mL, respectively, when supplemented with acetate and TCE, whereas supplementation with ethanol and TCE yielded bacterial populations of 1.39×10^7 and 1.39×10^6 MPN/mL, respectively.

Black spots appeared within the BART SRB reactors, indicating "very aggressive" SRB populations. The formation of dark yellow and brown opacity and precipitate, gas bubbles, and slime plaques on the IRB BART indicator balls indicated that the monitoring well MW-29 and MW-31 samples contained "very aggressive" and "moderately aggressive" IRB populations, respectively.

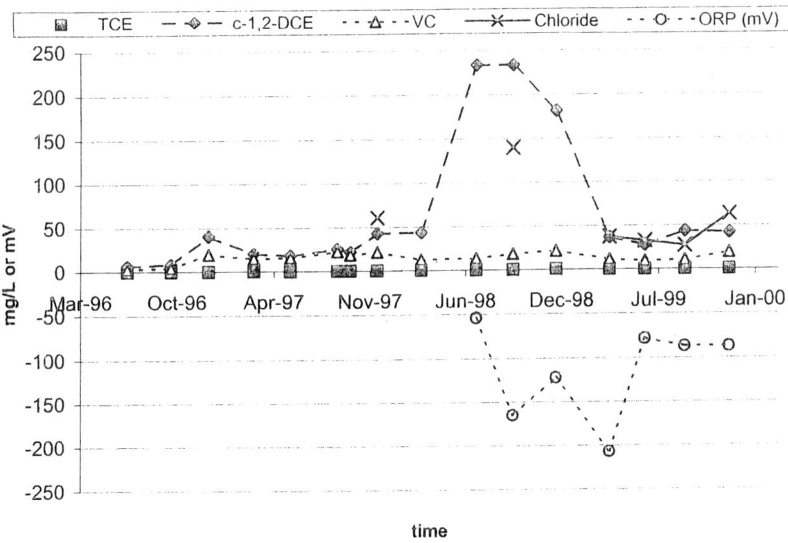

FIGURE 3. Chlorinated ethene and chloride concentrations (mg/L), and ORP (mV) versus time in groundwater at monitoring well MW-31.

TABLE 2. Summary of laboratory TCE degradation test.

All results in mg/L, from duplicate samples.

	Day 0			Day 9			Day 16		
Condition	**TCE**	**c-1,2-DCE**	**VC**	**TCE**	**c-1,2-DCE**	**VC**	**TCE**	**c-1,2-DCE**	**VC**
An. Native	5.80	**.58 U**	.05 U	.05 U	4.85	.05 U	.05 U	4.80	.05 U
An. Killed	6.15	.05 U	.05 U	6.10	.05 U	.05 U	6.25	<50	.05 U
An. EtOH-Amended	6.70	.05 U	.05 U	3.10	2.90	.05 U	3.30	2.90	.05 U
An. EtOH/ SO_4^{2-} - Amended	6.15	.05 U	.05 U	.05 U	5.20	.05 U	.05 U	5.15	.05 U
Aer. Native	NT	NT	NT	NT	NT	NT	6.50	0.83	0.03 J

Bold: Indicates that only one value was above the detection limit, and that the detection limit value was averaged with the detected value.

Aer.: Aerobic
An.: Anaerobic
TCE: trichloroethene
U: Not detected at cited concentration
c-1,2-DCE: cis-1,2-dichloroethene
J: estimated value
VC: vinyl chloride
NT: Not tested

The results of the biotreatment study are summarized in Table 2. Sulfate amendment was examined in response to the "very aggressive" nature of SRB, as discerned during the BART test. Ethanol amendment was examined in response to the relatively high enrichment of anaerobes observed in the MPN counts where ethanol was added as a carbon source. The results indicate that most if not all of the TCE was transformed to *c*-1,2-DCE in the native anaerobic and anaerobic ethanol-/sulfate-amended batches by Day 9 of the study. Less transformation appeared to occur in the presence of ethanol alone, a result which cannot presently be explained. VC did not appear to be produced during the transformation processes. The ethanol-/sulfate-amended batches contained a black precipitate (presumably reduced sulfur) by Day 16, indicating that a highly anaerobic environment had formed. No loss of TCE was observed in the killed anaerobic control or native aerobic batches. The native aerobic batches contained a yellow precipitate (presumably oxidized iron) by Day 4.

CONCLUSIONS

The field data indicate that significant decreases in CE concentrations occur with distance along the groundwater flow path from the release source. The field data also indicate that significant reductive transformation of TCE to daughter products and chloride has occurred in the Overbank zone of the Site despite the high concentrations of CE constituents. The strength of the reducing environment is reflected in the low dissolved oxygen concentrations, the negative ORP readings, and the relatively high methane and Fe^{2+} concentrations. *C*-1,2-DCE and VC appear to degrade as well. The laboratory studies indicate that the Site groundwater contains significant and aggressive populations of the types of bacteria that are known to play a role in biological transformation of CEs, and that the Site groundwater can support a high rate of TCE transformation to *c*-1,2-DCE.

ACKNOWLEDGMENTS

The authors wish to thank: Eileen L. Webb and Karen F. Williams for field exploration efforts; Scott Kranz and James R. MacPherson, Ph.D., for hydrogeology and fate & transport interpretations; and Sean F. Gormley for analytical support.

REFERENCE

Wiedemeier, T.H., M.A. Swanson, D.E. Moutoux, E.K. Gordon, J.T. Wilson, B.H. Wilson, D.H. Kampbell, P.E. Haas, R.N. Miller, J.E. Hansen, and F.H. Chapelle. 1998. Technical Protocol for Evaluating Natural Attenuation of Chlorinated Solvents in Ground Water. United States Environmental Protection Agency publication EPA/600/R-98/128, September 1998.

ASSESSING CONTAMINANT PLUME STATUS AT INTRINSIC REMEDIATION SITES

Travis Williamson, Mark Kelley, (Battelle Memorial Institute, Columbus, Ohio), Arthur Fisher, Rajagopal Krishnamoorthy (Naval Air Station Fallon, Fallon, Nevada)

ABSTRACT: A two year feasibility study was conducted at Naval Air Station (NAS) Fallon, Nevada to evaluate the viability of intrinsic remediation for groundwater contaminated with chloroethenes and fuel hydrocarbons at seven Installation Restoration (IR) sites. An important element of the feasibility study is the determination of the plume status at each site, because it provides indirect evidence for intrinsic remediation of the contaminants. However, plume status is not easily determined. Methods that have been suggested for assessing changes in the status of a plume include: simple visual analysis of contaminant concentration vs. distance plots using data from wells located along the plume centerline and collected over several monitoring events (referred to in this paper as the spatial method); temporal (i.e., trend) analysis of contaminant data from individual monitoring wells located throughout the plume; and construction of contaminant concentration contour maps for multiple sampling events (American Society for Testing and Materials, 1999; USEPA, 1992; USEPA, 1999; Wiedemeier et al., 1996). These methods often provide inconclusive results. A revised statistical approach (Naber et al., 1997) that combines both temporal (i.e., time series) and spatial (i.e., multiple monitoring well locations) data into a single analysis was used to better elucidate trends and make a final assessment of plume status.

INTRODUCTION

A comprehensive groundwater monitoring program was implemented at NAS Fallon as part of the intrinsic remediation feasibility study. Historical monitoring data collected before the feasibility study dates back to 1991, however, these data were considered to be too sparse for the purpose of assessing plume status. Therefore, five semiannual sampling events were conducted during the two-year intrinsic remediation feasibility study beginning in March 1997. Groundwater parameters that were measured during this period included the contaminants of interest at each site and geochemical parameters that can provide evidence for biodegradation of the contaminants (dissolved oxygen, nitrate, sulfate, ferrous iron, manganese, methane, hydrogen, and redox potential). This paper focuses only on the analysis of contaminant data.

Objective. The objective of this paper is to illustrate the usefulness of a new statistical method (the combined spatial/temporal method) for assessing plume status, by comparing results obtained with this method to two other commonly used methods for assessing plume status: temporal (i.e., trend) analysis of contaminant data from individual monitoring wells located throughout the plume; and, simple visual analysis of contaminant concentration versus distance plots using data from wells located along the plume centerline and collected over several monitoring events. A third method that is often used involves the construction of contaminant concentration contour maps for multiple sampling events. However, this method was considered to be infeasible at NAS Fallon due to the excessive analytical costs that would be incurred.

Site Description. This paper presents the assessment of plume status for three contaminants at Site 1, which is one of the seven IR sites at NAS Fallon included in the intrinsic remediation feasibility study. Site 1 was a former Crash Crew Fire Training Area that was used from the mid-1950s until April 1988. During the time period that the site was operated, approximately 1.1 million gallons of flammable liquids were burned at the site including waste products from the base fuel farm, aircraft maintenance, and vehicle maintenance (Naval Energy and Environmental Support Activity, 1984). Groundwater at the site is contaminated with the chloroethenes tetrachloroethene [PCE], trichlorethene [TCE], *cis*-1,2-dichloroethene [*c*DCE], *trans*-1,2-dichloroethene [*t*DCE], 1,1-dichloroethene, and vinyl chloride and dissolved fuel hydrocarbons including benzene. Figure 1 shows the extent of groundwater that is contaminated with PCE, TCE, and *c*DCE, the primary chloroethenes at the site, and benzene, as determined from data collected during the first two semi-annual sampling events (March and September 1997). Because of the large number of wells that were required to construct the contour maps, contour maps were not constructed thereafter and all subsequent sampling events focused on sampling only wells located along a transect through the plume center.

METHODS FOR ASSESSING PLUME STATUS

Spatial Analysis. The spatial analysis involved constructing a concentration versus distance plot for each contaminant using data from monitor wells located along the plume centerline and collected over multiple sampling events. The plot is constructed with its origin located at the source and aligned in the direction of groundwater flow. Thus, the distance at which the concentration curve intercepts the x-axis is an estimate of the location of the leading edge of the plume. Furthermore, because this plot combines both spatial and temporal data, trends in concentrations at points along the transect can be inferred from upward or downward changes in the concentration curve over time. Changes in the position of the plume leading edge over time can be detected with this method only if the plume advances or retreats a sufficient distance such that contamination either is detected for the time in a downgradient well (plume advancement), or, contaminant levels in the most-downgradient well with a history of contamination drop below detectable levels (plume retreat). Another disadvantage of this method is that it requires a consistent set of monitoring data for the transect wells. At NAS Fallon, where new wells were installed along the plume transects in 1997, this meant that historical monitoring data collected prior to 1997 could not be included in the assessment.

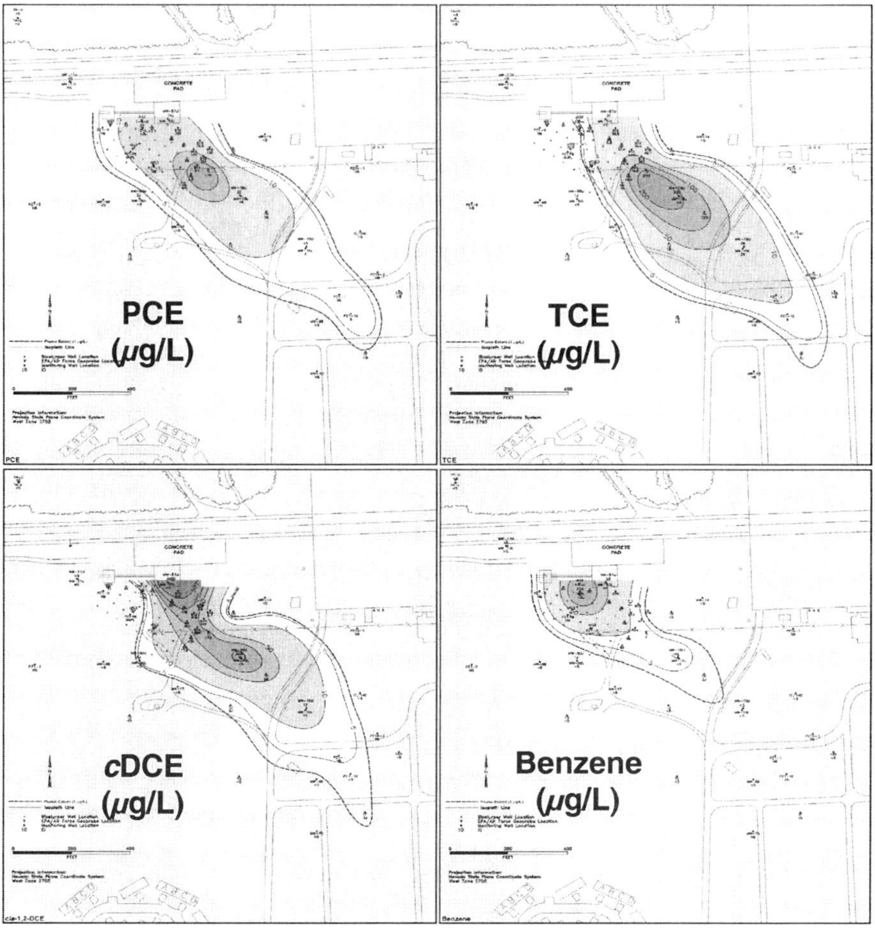

FIGURE 1. Map of NAS Fallon Site 1 showing the extent of PCE, TCE, *c*DCE, and benzene in groundwater

Temporal Analysis. The temporal (i.e. trend) analysis involved performing a linear regression of time-series concentration data for individual monitoring wells located throughout the plume. The method uses data from all of the wells at the site to estimate the variability in the data for a given contaminant, then uses this variance estimate to assess the significance of the trend for each individual well. Trends were considered significant if the p-value was < 0.05; p-values greater than 0.05 meant that there is insufficient data to conclude that the slope is different from zero.

The results of the trend analysis were plotted on the plume maps to evaluate the overall status of each plume. If contaminant concentrations were increasing in all of the wells at the site or in the wells at the leading edge of the plume, the plume was considered to be expanding. Similarly, if concentrations appeared stable (i.e., neither significantly increasing or decreasing) in most wells, then the plume status was considered stable; and, if concentrations were decreasing in the majority of wells, the plume status was considered shrinking.

Combined Spatial/Temporal Analysis. The statistical method developed by Naber et al. (1997) was applied at NAS Fallon in an attempt to assess plume status. This method incorporates all available monitoring data for all wells at a site to determine an overall trend for the plume. The method arises from that presented by Buscheck and Alcantar (1995) which assumes that the contaminants exhibit first-order decay. Concentration trends are first assessed on a well by well basis, using the same techniques as those used in the temporal analysis discussed above, and then this information is combined to determine an overall trend for the entire site. A monitoring well is included in the analysis of a contaminant only if there were two or more detections of the contaminant over time, therefore the analysis is not biased by including wells that are located outside the area of contamination.

RESULTS AND DISCUSSION

The three methods for evaluating plume status (spatial, temporal, and combined spatial/temporal) gave varying and sometimes inconclusive results regarding plume status at Site 1, but there are some results that were consistent across the various methods. The following subsections present the analyses and results for PCE, *c*DCE, and benzene at Site 1.

Spatial Analysis. Figure 2 shows the concentrations of PCE, *c*DCE, and benzene measured during each semiannual sampling event with the x-axis beginning in the source area (i.e., the former fire training pit) and extending down gradient in the direction of groundwater flow.

As Figure 2 shows, some noticeable changes exist within the contaminant data, but variability in the results along the plume transect makes it difficult to evaluate plume status with any degree of certainty. PCE appears to be decreasing in wells MW-51U and 1I, which are located in the source area, and increasing in MW-16U where the plume center of mass is located. Concentrations in down gradient wells MW-15L, PZ1-10, and 1F are stable to increasing, suggesting that the PCE plume may be expanding.

In contrast to PCE, *c*DCE concentrations appear stable in each well sampled along the plume transect with the exception of MW-51U which shows decreasing concentrations over time in the source area, suggesting the plume is either stable or slightly shrinking.

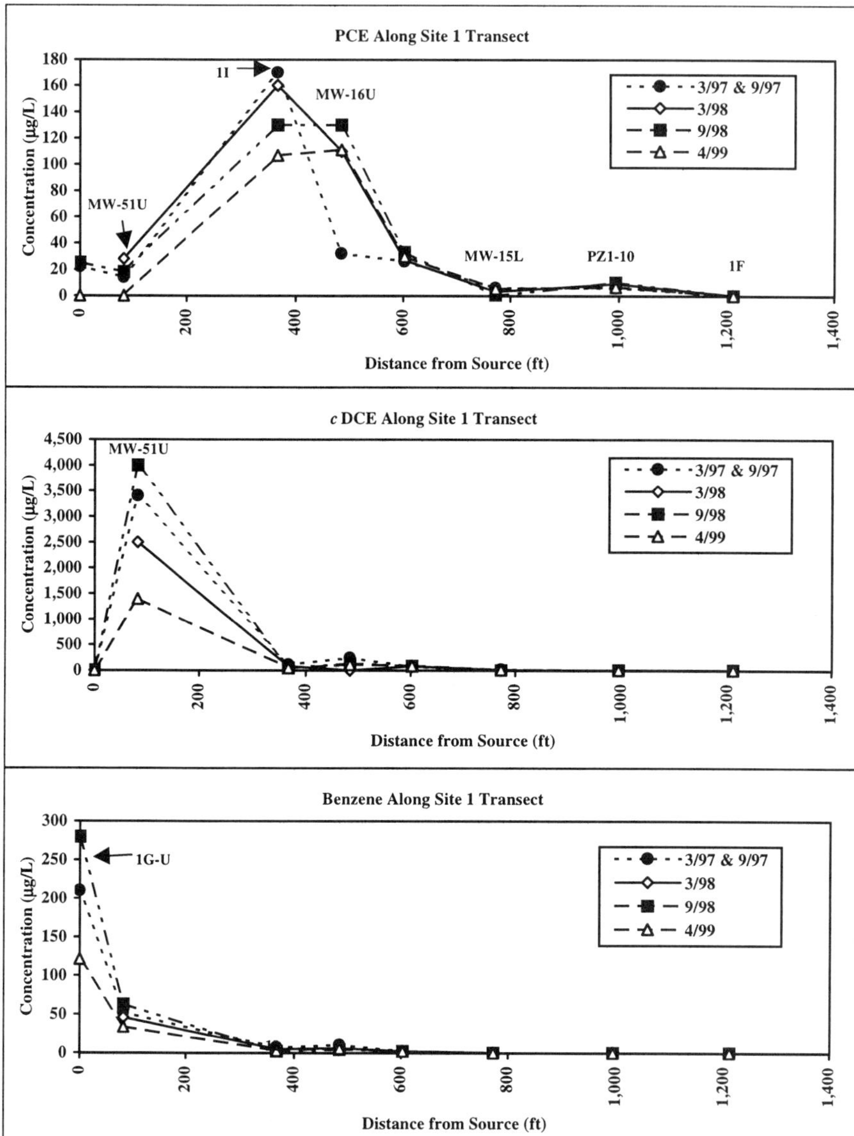

FIGURE 2. Concentration vs. distance along the plume centerline plots for PCE, *c*DCE, and Benzene for the spatial method.

Trends of benzene concentrations are more difficult to infer due to a large amount of variability in data near the source area (well 1G-U). Overall, the majority of benzene concentrations are stable to decreasing, suggesting the benzene plume is either stable or shrinking.

In summary, general changes in contaminant concentrations can be identified from transect plots showing results from multiple sampling events, but a final determination of plume status is largely a matter of judgement and is often difficult to make when concentrations do not change consistently along the plume transect.

Temporal (Trend) Analysis. Individual trends determined with the temporal analysis can be displayed on a plume map to assess important changes that may be occurring in the plume in order to make a determination of plume status (Figure 3). A significant decreasing trend is indicated by a negative slope with a p-value <0.05; whereas, a significant increasing trend is indicated by a positive slope with a p-value <0.05. Figure 3 displays the significant trends that were calculated with the temporal analysis for PCE. PCE is stable in the source area well MW-51U while increasing in well MW-16U located near the center of the site. A significant increasing trend of PCE exists in source area well MW-51L; however, because the well is screened in the lower part of the aquifer, this result was not considered in the assessment of PCE plume status. The interpretation of individual trends for PCE shown in Figure 3 leads to the conclusion that the PCE source is becoming depleted and contaminated groundwater is migrating as a "slug" of contamination toward well MW-16U; *c*DCE is shown in Figure 3 to be stable in all but one well, suggesting a stable plume. Significant decreasing trends of benzene were detected in several wells throughout the plume, suggesting that the plume is shrinking over time. The temporal analysis is more robust than the spatial analysis (i.e., transect plots) for assessing plume status at a site because it is able to incorporate statistical analysis of contaminant data to identify trends, but the final determination of plume status is still largely dependent on judgement.

Combined Spatial/Temporal Analysis. With this method, the linear regression equation used to model the changes in contaminant concentration over time in individual wells is combined into a single equation that estimates overall contaminant trend and p-value of that trend at a site. The calculated trends and p-values for PCE, *c*DCE, and benzene at Site 1 are as follows: 0.112 (<.0001), -0.019 (0.0921), and –0.018 (0.0001), respectively. Thus, the method suggests that PCE has a statistically significant increasing trend, *c*DCE does not have a statistically significant trend, and benzene has a statistically significant decreasing trend. These results suggest that the PCE plume is expanding, the *c*DCE plume is stable, and the benzene plume is shrinking, which is consistent with the conclusions made by the spatial and temporal analyses. However, in its current form, this method is unable to distinguish that the increasing trend in PCE is largely due to downward migration of contamination in the source area (as evidenced by the significant increasing trend in well MW-51L) and a downgradient "shifting" of the plume center of mass rather than expansion of the plume at its leading edge. Thus, more work is required to make the method more "intelligent".

The combined spatial/temporal analysis provides statistically defensible estimates of contaminant plume status at a site by using all available monitoring data and the method is more robust than both the spatial and temporal analyses methods because determination of overall plume status is less dependent on judgement

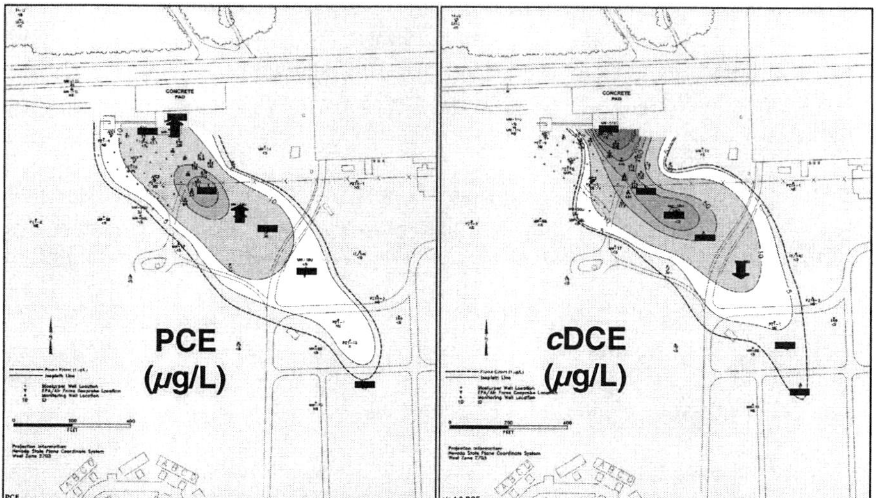

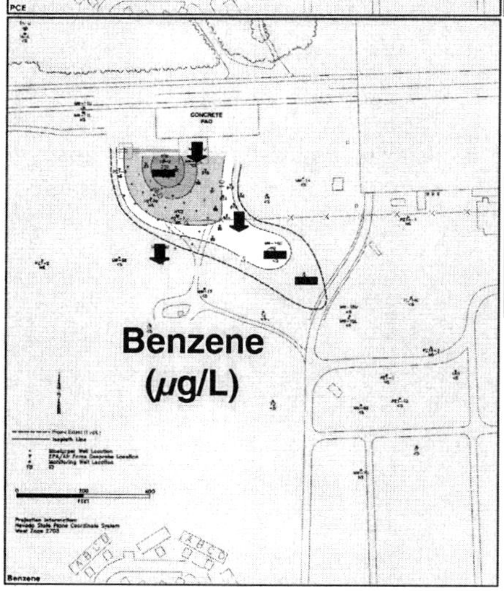

FIGURE 3. Temporal analysis results for PCE, *c*DCE, and benzene at NAS Fallon Site 1.

CONCLUSIONS

Assessing the status of a contaminant plume is required to determine the effectiveness of intrinsic remediation. Commonly used methods for determining the status of a plume sometimes fail to provide a clear indication of important trends that exist at a site because they rely heavily on the judgement of the interpreter. The combined spatial/temporal analysis that is being used at NAS Fallon allows for significant trends to be identified without relying on judgement.

REFERENCES

American Society for Testing and Materials. 1999. *Standard Guide for Remediation of Ground Water by Natural Attenuation at Petroleum Release Sites.* E1943-98. ASTM, West Conshohocken, PA.

Buscheck, T.E. and C.M. Alcantar. 1995. "Regression Techniques and Analytical Solutions to Demonstrate Intrinsic Bioremediation," In R.E. Hinchee, J.T. Wilson, and D.C. Downey (Eds.), *Intrinsic Remediation*, pp. 109-116, Battelle Press, Columbus, OH.

Naber, S., Buxton, B., McMillan, M, and Soares, A. 1997. "Statistical Methods for Assessing the Effectiveness of Intrinsic Remediation," In *Proceedings of the Fourth In Situ and On-Site Bioremediation Conference*, Battelle Press, Columbus, OH.

Naval Energy and Environmental Support Activity. 1984. *Naval Installation Restoration Program Management Guidance.* NEESA 20.2-047B. Defense Technical Information Center, Port Hueneme, CA.

U.S. Environmental Protection Agency. 1992. *Methods for Evaluation the Attainment of Cleanup Standards, Volume 2: Groundwater.* EPA/230/R-92/014. Office of Solid Waste and Emergency Response, Washington, DC.

U.S. Environmental Protection Agency. 1999. *Use of Monitored Natural Attenuation at Superfund, RCRA Corrective Action, and Underground Storage Tank Sites.* Directive 9200.4-17P. EPA/50/R-99/009. Office of Solid Waste and Emergency Response, Washington, DC.

Wiedemeier, T., M.A. Swanson, D.E. Moutoux, E.K. Gordon, J.T. Wilson, B.H. Wilson, D.H. Kampbell, J.E. Hansen, and P.H. Haas. 1996. *Technical Protocol for Evaluating Natural Attenuation of Chlorinated Solvents in Groundwater.* Air Force Center for Environmental Excellence, Technology Transfer Division, Brooks Air Force Base, San Antonio, TX.

NATURAL ATTENUATION OF CHLORINATED SOLVENTS IN GROUNDWATER - BIS(2-CHLOROETHYL)ETHER ANAEROBIC TRANSFORMATION

A. Basel Al-Yousfi, J.K. Gotpagar, K.C. Lee, and Sunil I. Shah
Union Carbide Corporation
South Charleston, West Virginia, USA

ABSTRACT: Overlapping organic contaminant plumes underneath a closed chemical manufacturing unit, were monitored to evaluate the intrinsic cleanup occurring in the groundwater. The field observations have indicated that BCEE ($C_4H_8Cl_2O$) has been transforming naturally in groundwater into Diethylene Ether (DEE, $C_4H_8O_2$) under the existing highly reducing conditions. Anthropogenic hydrocarbon compounds (Benzene, Toluene, Ethylbenzene, and Styrene, BTES) present under the site are acting as electron donors, and facilitating DEE production. The Natural Attenuation monitoring program has been implemented semi-annually, and will continue in the future to ensure adequate reduction of groundwater contamination at the site.

INTRODUCTION

Traditionally, the compounds thought to be amenable to natural attenuation were only considered to be petroleum hydrocarbons like BTEX. Chlorinated compounds were viewed as recalcitrant and hence not readily bidegradable. However, more and more field experience and research carried out in the past decade have shown promising potential of biodegradation as a viable alternative for remediation of sites contaminated with chlorinated compounds. Given the appropriate hydrogeologic and redox conditions, it is possible to achieve intrinsic bioremediation of the site. Current paper deals with one such industrial site in Texas, where Monitored Natural Attenuation (MNA) is successfully being implemented. Historical data sets at this site have captured considerable reduction in the plume sizes and the corresponding accumulation of daughter products.

Site Background. The subject site is located in a chemical manufacturing plant, which has been operated since 1954. Some of the old production units, however, were shut down in early 1980's. Among those included Olefins 1, which is of special interest concerning this groundwater natural attenuation monitoring work, that was started in early 1996. The plant is located on the Coastal Plain that slopes gently towards the Gulf of Mexico with a gradient of approximately two to five feet per mile. Regional ground water flow is toward west to Mission Lake and toward south to San Antonio Bay. Groundwater flow in the Central Plant is under confined conditions, with hydraulic gradients beneath the production facilities around 0.00004 ft/ft. The average hydraulic conductivity along the flow from contaminant source area to plant boundary is approximately 0.00167 cm/sec, and the average groundwater velocity in the Central Plant area is 10 ft/yr (RFI Task II, UCC, 1995).

AREA OF NATURAL ATTENUATION - OLEFINS 1

Olefins 1 area was operated between 1954 and 1983. There has been no manufacturing activity in this area since 1983. Ground water sampling results indicated the presence of three major contamination plumes, Benzene (BTEX or BTES), Bis(2-Chloroethyl) Ether (BCEE), and Diethylene ether (DEE). The major area of BTES contamination is located beneath the former gas compressor building, near Well C-88-66A (Figure 1), where an oil pool has been reported in the past at the northeast corner of the block. BTES and other polynuclear aromatic hydrocarbons (PAHs) were by-products of the olefins cracking furnace operation that produced ethylene gas. The main BCEE contamination plume in the area coincides with the southernmost end of the Olefins 1 refinery operation. BCEE was used as an extracting solvent to refine butadiene for 20 years in the former butadiene refinery unit; and was last used in the unit in 1973. The Butadiene Refinery Unit was the only operation that used BCEE in the plant. The center of the BCEE plume was predicted to have migrated 200 to 400 feet northward in a period of 20 to 40 years. Groundwater monitoring over past few years have revealed that the BCEE plume has stabilized and is shrinking. DEE has never been handled in this particular unit. Moreover, on the upgradient side is an above ground cooling pond and there is no operating unit. Hence, DEE is likely transformed in this area through dechlorination of BCEE (One of the two major manufacturing processes for DEE is to react BCEE with water at elevated temperatures under caustic condition.). Over 15 monitoring wells have been used in the area for groundwater sampling in support of this natural attenuation demonstration project (Figure 1).

Analytical results from the monitoring wells have been used to complete the delineation of affected areas, to understand the fate and transport characteristics of the plumes, and to track bioremediation cleanup processes. Figure 1 shows the location of wells used in this study. Release of BCEE occurred approximately 25 to 45 years ago near Well 286 at the old butadiene unit. Currently, the center of BCEE plume has migrated to Wells C-93-198A, C-93-192A and C-88-66A area.

NATURAL ATTENUATION MONITORING PROGRAM

Groundwater sampling events and chemical analyses have been routinely performed to provide affirmative indicators of the ongoing intrinsic bioremediation. The groundwater sampling and analyses are conducted on semi-annual basis since early 1996. Analytical results have been employed to determine the distribution of dissolved phase contaminants (BTES, BCEE and DEE), and all of the parameters pertinent to documenting natural attenuation (aerobic and/or anaerobic). The analytical parameters involve groundwater plume constituents, electron acceptors, and metabolic byproducts. A portable apparatus unit was utilized to facilitate the *in situ* measurement of pH, Oxidation Reduction Potential (ORP), and Dissolved Oxygen (DO) in the field. Some chemical analyses (i.e., geochemical indicators) were conducted in internal laboratories , and others (i.e., volatile and semi-volatile constituents) were done in a certified, commercial laboratory.

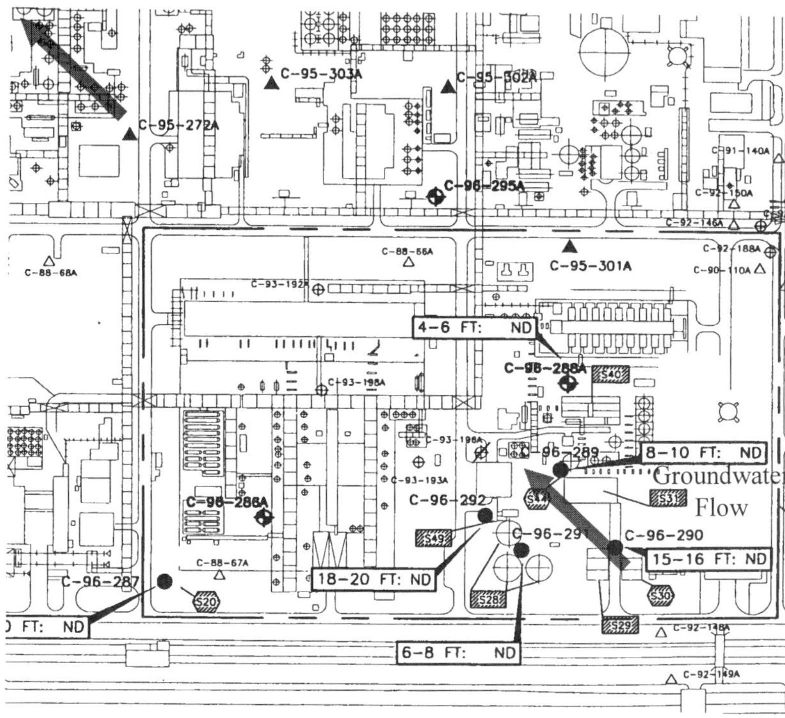

FIGURE 1. Location of Different Wells in Olefins 1 Area

RESULTS OF MONITORED NATURAL ATTENUATION

Geochemical Indicators. The dissolved oxygen (D.O.) contents in the groundwater varied from 1.3 mg/l (upgradient) to about 0.4 mg/l (or equivalent to zero D.O.). Relatively lower dissolved oxygen concentrations were observed for wells located near the center of the plume: Wells C-88-66A, C-93-192A, and C-93-198A (Figure 1). Nitrate concentrations in most groundwater wells were non-detect, indicating that nitrate (NO_3^-) may not occur naturally in this groundwater aquifer. This observation was also confirmed in the earlier data sets.

Upgradient and downgradient from the plume center, one can depict the highest groundwater sulfate concentrations in the immediately adjacent area at 1,500 to 2,600 mg/l, respectively. It should be noted that data in the other areas beneath the plant show sulfate concentrations as high as 6000 to 7000 mg/L. Lower concentrations observed in the adjacent area to Olefins-I, could be attributed to the dilution effect caused by the water mounding underneath the cooling water basin. Within the plume's core however, sulfate concentrations drop considerably to as low as non-detect (ND) in Well C-88-66A, which cannot be explained by dilution phenomenon alone. This indicates that significant sulfate reduction via anaerobic biological reactions have been occurring in the Olefins 1 Area (Figure 2). This is further confirmed by the variation in TOC and sulfate contents in different wells in the Olefins-I area. Figure 2 shows that TOC concentrations reversibly mirror those of sulfate, so that higher groundwater contamination coincides with relatively lower values of sulfate (and vise versa).

Other supporting data collected on the degradation products also reveal the same. For example methane contents varied from zero concentration in the background wells to maximum of 17 mg/L within the core plume wells (not shown). The high concentrations of methane observed were consistent with the relatively lower D.O., and higher concentrations of TOC and BTES within the core plume area, indicating the prevailing conditions for anaerobic bioremediation. Furthermore, dissolved CO_2, and free chloride ions (Cl^-) were also significantly higher (> two times the background values) in core plume wells. This was reflected in Wells C-88-66A, C-93-192A, and C-93-198A (all are in the center of the affected area). Furthermore, some BCEE was also found around these wells. Therefore, anaerobic biological activities (reductive dehalogenation under methanogenesis and sulfate reduction conditions) are likely occurring at the core of the groundwater plume in the Olefins 1 Area.

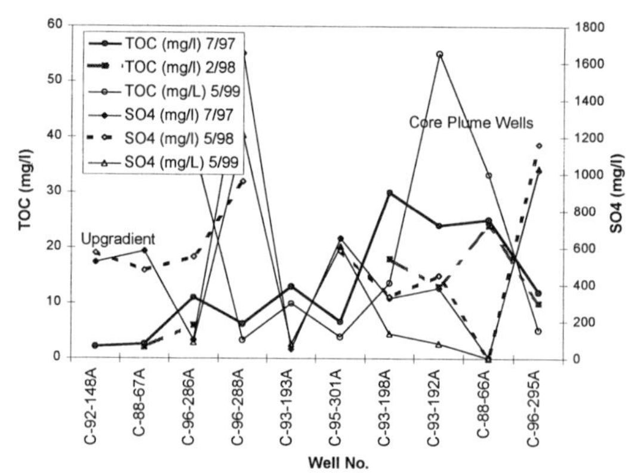

FIGURE 2. Variation in TOC and SO_4^{2-} with Location of the Well

Contaminant Reduction. Data on TOC temporal trend showed definite decrease in organic contents, especially in the vicinity of the plume's core (Figure 2). An overall reduction in TOC concentrations near the center was estimated at 30 to 65% between January 1997 and May 1999. Furthermore, the redox measurements revealed that the higher TOC concentrations occurring at the center coincide clearly with negative ORP values, indicative of anaerobic biodegradation processes under reduced conditions. The historical groundwater contamination data for the Olefins-1 area have shown continuous reduction in both BTES and BCEE contamination, especially in the core of the plume. It is of importance to stress that both Benzene and BCEE plumes are highly soluble and mobile, and therefore less susceptible to adsorption. Furthermore, due to the confined characteristics of the ground water aquifer, volatilization plays minimal role in extracting contaminants from the aqueous phase. Also, due to the low groundwater velocity under the site, dilution and migration (wash out) have minimal impacts on reducing contaminant concentrations. In addition, BTES compounds are readily amenable to natural intrinsic bioremediation under both aerobic and anaerobic conditions. Therefore, majority of COCs removal from the ground water plume could be directly related with the subsurface intrinsic bioremediation.

HISTORICAL DECLINE IN BCEE CONCENTRATIONS/MASS AND CORRESPONDING DEE ACCUMULATION.

A comparison with older BCEE groundwater data collected at the site, in the period between 1993 and 1995, revealed that significant reduction in BCEE concentration has occurred in the 1996-1997 time-frame for all the BCEE containing wells in the area. This sudden drop in BCEE concentration can be explained by the finally depleted BCEE groundwater contamination source release underneath the former butadiene unit, where BCEE was used. No BCEE has been used in the area since 1973. Monitoring data indicated that after approximately 25 years, continuous but passive bleeding into groundwater may have eventually stopped. Other hypotheses, such as a sink (e.g., sewer) which drew up groundwater contamination quickly, cannot be postulated because other COC(s) in the area (e.g., BTES) have not shown any similar steep reduction in concentration during the same period (or any other period throughout the monitoring program). Again, although the concentration profiles are declining with time, the highest measured concentrations were currently found in Wells C-88-66A, C-93-192A, and (to a lesser extent) C-93-198A.

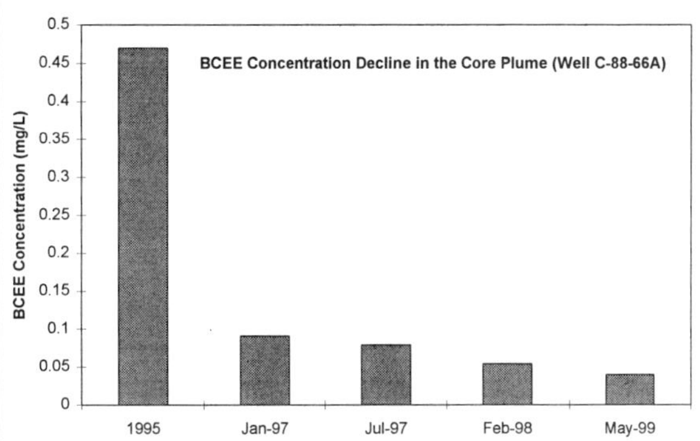

FIGURE 3. BCEE Concentration Decline with time in the Well located in the Core of the plume

Figure 3 plots the decline in BCEE concentration in core well, C-88-66A with time. It should be noted that many wells on the plume boundary have been consistently non-detect for BCEE, indicating that the decline in BCEE in core wells is not due to the migration of the plume or its constituents. Furthermore, this decline is consistent with increase in the DEE groundwater concentrations. DEE was never used or manufactured in the Olefins 1 area. Serious effort to measure DEE concentration in the Olefins1 started in the Fall of 1996 due to suspicion that a DEE plume may exist as the result of degradation product from BCEE. Data collected since then have shown that slightly higher concentrations of DEE occur slightly downgradient from the plume center as defined by BCEE concentrations. No prominent trend of DEE concentration reduction has been observed during the brief (3 years) monitoring period so far. This could be due to the relatively recalcitrant characteristics of DEE for biodegradation, the low dilution effects resulting from low groundwater velocity, and the continuous production of DEE as a daughter product of BCEE degradation. Comparison of BCEE ($C_4H_8Cl_2O$) plume and DEE ($C_4H_8O_2$) plume depicted in Figure 4 strongly suggests that there is a mutual relationship between the two compounds. It is believed the degradation process is a Type 1 chlorinated plume behavior (Wiedmeier et al., 1996). This hypothesis was further verified by comparing the data on other geochemical indicators and daughter product of degradation of BCEE, chloride ions. Within the center of the plume SO_4^{-2} concentration was at its lowest value (Figure 2), indicating the sulfate mediating redox conditions, and Cl^- concentration is at its highest (Figure 4, indicating dechlorination process). Accordingly, the plume core has been undergoing strong reductive dehalogenation of BCEE (electron acceptors) in the presence of high contents of BTES organics and sulfate (as electron donors). Figure 4 further shows that DEE peaks occur immediately after BCEE peaks, which strongly suggests the hypothesis that DEE is a byproduct of BCEE biodegradation in ground water (Wiedmeier et al., 1996). It might be of importance to reiterate that the center of the BCEE groundwater plume in the area has migrated 200 to 400 ft northward in the last 20 to 40 years - this may provide an explanation to the presence of chloride and DEE upgradient of the current plume. Other potential intermediates and daughter prod-

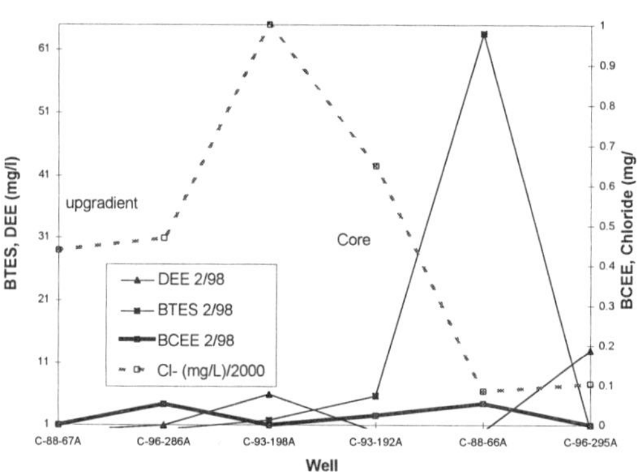

FIGURE 4. Relationship between BTES/BCEE degradation, and Cl⁻/DEE formation

ucts (e.g., Vinyl Chloride, VC) that may result from the processes of dechlorination of certain chlorinated solvents were analyzed for, and found to be non-detect in all of the monitoring wells.

As referred to earlier, groundwater data indicated that passive depleting source(s) of BCEE most likely has stopped after 1996. All data observed after 1996 therefore could be the result of degradation of the fixed initial mass in the area that is associated with the concentration observed in 1996. Based on this assumption, BCEE and DEE data obtained after 1996 can be used to calculate first order reaction kinetic rate constants for the degradation of BCEE into DEE, and from DEE into end products as the two first-order consecutive reactions shown as follows (Smith, J.M., 1981):

$$A \xrightarrow{K2} B \xrightarrow{K3} C$$

Above kinetic model was applied to simulate BCEE and DEE historical data in Wells C-93-192A and C-93-198A in the core of the Olefins 1 plume, with the initial conditions as: [B] = [C]=0 and [A]=[A_0] at t=0.

Comparisons between predicted and actual ground water concentrations were carried out and is presented in Figure 5. Figure 5 shows that reasonable agreement between the simulated and observed field results was obtained. Regression analyses were also performed to predict the first-order rate constants (K_2 and K_3) and half-life parameters concerning BCEE and DEE decomposition. For BCEE, the first-order kinetic rate constant was estimated between 1.35 and 1.46 [$year^{-1}$], resulting in a half-life degradation value of 0.5 year. For DEE, the first-order kinetic rate constant was estimated between 0.25 and 0.74 [1/ year], resulting in half-life degradation values varying from 1 to 3 year. These values compare reasonably well with those reported in the literature (Howard et al, 1991).

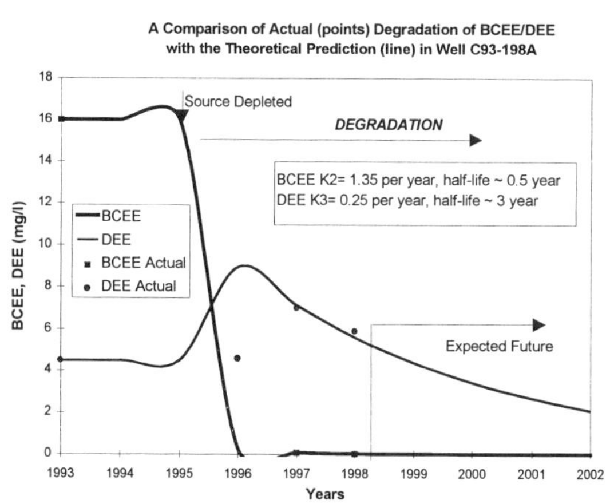

FIGURE 5. Comparison between the observed and predicted BCEE/DEE profiles

CONCLUSIONS

In summary, organic contaminants plumes have stabilized, shrinking, and are essentially confined within the Olefins 1 block. Multiple evidences support the notion of natural attenuation cleanup under the site. These evidences have been documented and ranked based on widely accepted NA protocol to provide tangible and quantified demonstration of the intrinsic bioremediation process. Although, the weight of evidence scoring results classifies the site as having "adequate evidence for biodegradation of chlorinated organics", spatial trend of geochemical indicators, combined with the very significant (1,500-2,600 mg/L) reduction in sulfate ion, and the accumulation of daughter products (chloride ion, carbon dioxide, and methane) give strong evidence that the intrinsic remediation via natural cleanup is proceeding at reasonably fast pace beneath the Olefins 1 area. The ongoing monitoring program is likely to continue for three to five years before reaching a decision that the groundwater plumes are being removed naturally, and/or are shrinking and retreating sufficiently.

REFERENCES

Howard, P. H. et al., "Handbook of Environmental Degradation Rates", Lewis Publishers, 1991.

RFI TASK II Work Plan, Seadrift Central Plant Area, Union Carbide Corporation, August 28,1995.

Smith, J. M., "Chemical Engineering Kinetics", 3rd Edition, McGraw-Hill, Inc., 1981.

Wiedmeier, T. H., Wilson, J. T., and Hansen, J. E. et al., "Overview of the Technical Protocol for Natural Attenuation of Chlorinated Aliphatic Hydrocarbons in Ground Water Under Development for the U.S. Air Force Center for Environmental Excellence", Symposium on Natural Attenuation of Chlorinated Organics in Ground Water, EPA/540/R-96/509, ORD, 1996.

OBSERVED NATURAL ATTENUATION OF TCA IN GROUNDWATER

Daniel B. Stephens, Ph.D. and Nicole T. Nelson, Ph.D.
(Daniel B. Stephens & Associates, Inc., Albuquerque, New Mexico USA)

ABSTRACT: A case study is presented for an arid zone site where chlorinated solvents created a contaminant plume nearly 1 mile wide and 5 miles long. Although trichloroethene (TCE) is the most abundant chemical in the plume, 1,1,1-trichloroethane (TCA) usage surpassed that of TCE in the mid-1970s. A few years later, redesign of the wastewater system substantially reduced the TCA loading to groundwater. Field data show the rapid decay of the TCA pulse over about a 5-year period under natural conditions. This study used the field data to calculate the half-life of TCA and then compared the results to half-lives determined in laboratory experiments. For this site, the field and laboratory results were consistent.

INTRODUCTION

The chemical TCA, which is used in large quantities for degreasing, is one of several chlorinated solvents that are commonly found dissolved in groundwater beneath industrial sites throughout the U.S. TCA came into use primarily in the early 1970s in response to regulatory pressure to reduce or eliminate the use of other solvents, primarily TCE.

Under some conditions, concentrations of chlorinated compounds may decrease in groundwater without active remediation, due to both biological and non-biological processes. In fact, natural attenuation of chlorinated dissolved-phase organic compounds is often considered a preferred remedial alternative for many impacted sites. Natural attenuation can occur as a result of several processes such as hydrodynamic dispersion, dilution, volatilization, sorption, biodegradation, and abiotic degradation.

Under anaerobic conditions, TCA in groundwater can break down biologically in a sequence of reductive dehalogenation steps, to 1,1-dichloroethane and then to chloroethane. Abiotic processes facilitate further transformation of chloroethane to ethanol, which in turn is biologically transformed to carbon dioxide (Vogel and McCarty, 1987).

TCA also naturally attenuates abiotically into 1,1-dichloroethene (DCE) and acetic acid, which apparently are produced simultaneously in the process. Both compounds can be transformed under methanogenic conditions by microbes, with the DCE converting to vinyl chloride and acetic acid converting to carbon dioxide and methane (Vogel and McCarty, 1987).

DCE is one of the five most frequently detected volatile organic compounds in domestic water supplies, other than trihalomethanes (Cline and Delfino, 1989). However, DCE is not known to have been used as a primary solvent in industrial processes. Therefore, its presence in large quantities in groundwater, often in association with TCA, suggests that the DCE in groundwater is derived primarily from the abiotic degradation of TCA.

The appearance of DCE in groundwater is important because it apparently is much more toxic than TCA. Unless methanogenic conditions exist, the DCE will not likely degrade and will therefore persist in the subsurface. Cline and Delfino (1989) identified four sites in Arizona with DCE contamination, even though DCE had not been used at the sites. The DCE appeared to be a result of the degradation of TCA used in the industrial processes. The DCE concentrations ranged from 206 to 10,400 micrograms per liter (µg/L) and in each case exceeded concentrations of its parent compound TCA.

The rate of natural attenuation of TCA to DCE by abiotic transformation is an important environmental issue. For example, insight into TCA's degradation to DCE would be valuable for designing remediation strategies and predicting cleanup times. This knowledge could also be used in forensic analyses conducted to determine responsibility for cleanup costs. One way to characterize the degradation rate is through quantification of its half-life, which is usually determined in the laboratory under controlled conditions. However, reliable analysis of the fate and transport of TCA in the subsurface depends on chemical properties that are relevant to field conditions, and few field studies have been conducted to quantify the degradation of chlorinated volatile organic compounds. The purpose of this paper is to compare relevant literature values for TCA's half-life, as derived from laboratory experiments, to a value based on analyzing a field-scale TCA plume in Arizona.

LABORATORY ANALYSES OF HALF-LIVES

Table 1 shows the degradation rate of TCA to DCE expressed as a half-life as measured in several studies. Also shown are the Arrhenius coefficients for the experiments, which allow extrapolation of the half-life determined under laboratory temperatures to other temperatures. The Arrhenius temperature dependence is an empirical relationship derived from the slope of lognormal (ln) k versus 1/T where k is the rate constant (half-life=0.69/k) and T is temperature. The slope of the line is equal to E/R and the intercept is ln A, where E is the activation energy of TCA, R is the gas constant, and A is the pre-exponential factor. Half-lives are shown at 25 degrees Celsius (°C).

FIELD SITE

The field site used in this study is located in the interior portion of a large alluvial groundwater basin in semiarid southern Arizona. The site is believed to be "Site 1" shown in Table 1 of Cline and Delfino (1989). The aquifer underlying the site consists predominantly of sand with variable portions of gravel, silt, and clay. The depth to the water table is approximately 100 feet, and the direction of groundwater flow is to the west-northwest with a gradient of 0.003. The hydraulic conductivity near the site is roughly 100 feet per day. Based on an effective porosity of 0.2, the local mean groundwater velocity is approximately 550 feet per year. Supply wells at the site may have influenced the direction and rate of groundwater flow in places until the wells were shut down in about 1979.

Groundwater temperature is approximately 25°C, and pH is approximately 7.5 (in the extraction wells).

TABLE 1. Half lives and Arrhenius parameters

	Arrhenius Parameters		Half Life (yrs)	
	Activation Energy (kJ/mol)	A factor (sec^{-1})	$t_{½}$ (20°C)	$t_{½}$ (25°C)
Cline and Delfino, 1989	119±3	2×10^{13}	2.0	0.85
Jeffers et al., 1989	116±2	4.93×10^{12}	---	1.1
Mabey et al., 1983	112	1.5×10^{12}	1.4	0.6
Vogel and McCarty, 1987	---	---	2.8-19.0	---
Haag and Mill, 1988	---	---	---	0.96
Dilling et al., 1975	---	---	---	0.5

Industrial activity began at the site in the 1950s, at which time TCE was the chlorinated solvent universally used for vapor degreasing. In about 1972 or 1973, TCA replaced TCE as the chemical of choice for degreasing. Wastewater from the manufacturing process was discharged to floor drains and then to unlined ditches and ponds for about 5 years, until the ponds were lined. Chemical wastes, including solvents, were also reportedly disposed into unlined pits adjacent to the plant near the ditches.

The first groundwater sampling for chlorinated volatile organic compounds in the vicinity of the unlined ponds, ditches, and pits occurred in 1981. Dozens of monitor wells were constructed to assess the nature and extent of contamination. TCE was the chemical detected in the highest concentrations, up to 16,700 µg/L. Based on this concentration and the proximity of the monitor well to the source areas, the TCE was likely present as a dense nonaqueous-phase liquid (DNAPL). TCE concentrations remained very high until pump and treat remediation was implemented in about 1987.

The only known samples of plant effluent, collected in 1978 shortly after the lined ponds were installed, showed TCA concentrations of 33,000 µg/L and no detectable DCE. During the early monitoring years (1981 to 1988), the highest TCA concentration in groundwater was 1,050 µg/L, detected near one of the monitor wells at the disposal pits. The TCA concentration decreased significantly during this time. Because there was no known primary use for, or purchases of DCE at this site, the DCE is believed to be derived largely from the degradation of TCA. The DCE plumes detected at the site have evolved quickly and seem to expand in places, likely in response to TCA abiotic transformation.

For this study, we used the TCA concentrations measured from 1982 to 1988 to determine the extent of the TCA plume beneath the field site and estimate its mass. The rate of decline of the plume mass over this period was then used to calculate the half-life of the TCA beneath the site.

RESULTS AND DISCUSSION

Figure 1 shows a series of contour maps depicting the shrinking TCA plume. Data shown are the maximum values reported each year for a well (Table 2). Strikingly, most of the TCA naturally attenuated before active remediation began in 1987 to remove the TCE, the solvent used more than 15 years earlier.

The maps of the TCA plume were contoured to determine the mass present in groundwater, assuming that the aquifer thickness was 90 feet and the porosity was 0.3. The mass calculated for each time period contoured is shown in the corner of each illustration in Figure 1.

The TCA plume mass was also plotted over time (Figure 2). From the slope of this line (–0.78), the half-life is calculated to be 0.9 year. The R squared for this linear analysis is 0.899.

The field result compares remarkably well to the laboratory analyses (Table 1), which reported half-lives for TCA at 25°C ranging from 0.5 to 1.1 years. Thus, it appears that the laboratory-derived degradation rates are representative of field conditions in this case.

Complexity in the groundwater flow field prevented the successful application of the age-dating method based on the TCA and DCE concentration ratio proposed by Smith et al. (1994).

CONCLUSION

The transformation of TCA to DCE in groundwater at an industrial site in southern Arizona occurs at approximately the same rate as predicted in the laboratory with reagent grade chemicals under controlled conditions. The laboratory half-life must be determined at the prevailing field temperature.

REFERENCES

Cline, P. V., and J. J. Delfino. 1989. "Transformation Kinetics of 1,1,1-Trichloroethane to the Stable Product 1,1-Dichloroethene." In R. A. Larson (Ed.), *Biohazards of Drinking Water Treatment,* pp. 47-56. Lewis Publishers, Inc., Chelsea, MI.

Dilling, W. L., N. B. Tefertiller, and G. J. Kallos. 1975. "Evaporation Rates and Reactivities of Methylene Chloride, Chloroform, 1,1,1-Trichloroethane, Trichloroethylene, Tetrachloroethylene, and Other Chlorinated Compounds in Dilute Aqueous Solutions." *Environmental Science & Technology.* 9:833-838.

Haag, W. R., and T. Mill. 1987. "Effect of Subsurface Sediments on Hydrolysis of Haloalkanes and Epoxides." *Environmental Science & Technology.* 22(6): 658-663.

Jeffers, P. M., L. M. Ward, L. M. Woytowitch, and N. L. Wolfe. 1989. "Homogeneous Hydrolysis Rate Constants for Selected Chlorinated Methanes, Ethanes, Ethenes, and Propanes." *Environmental Science & Technology.* 23(8): 965-969.

Mabey, W. R., V. Barich, and T. Mill. 1983. "Hydrolysis of Polychlorinated Ethanes. Presented, Division of Environmental Chemistry, 186th Annual Meeting of the American Chemical Society, Washington, D.C., September 1983.

Smith, J. S., L. Eng, and T. Mill. 1994. "The Determination of the Age of 1,1,1-Trichloroethane in Groundwater." Presented, 46th Annual Meeting, AAFS, San Antonio, Texas, February 14-19, 1994.

Vogel, T. M., and P. L. McCarty. 1987. "Rate of Abiotic Formation of 1,1-Dichloroethylene from 1,1,1-Trichloroethane in Groundwater." *Journal of Contaminant Hydrology*. 1(1987): 299-308.

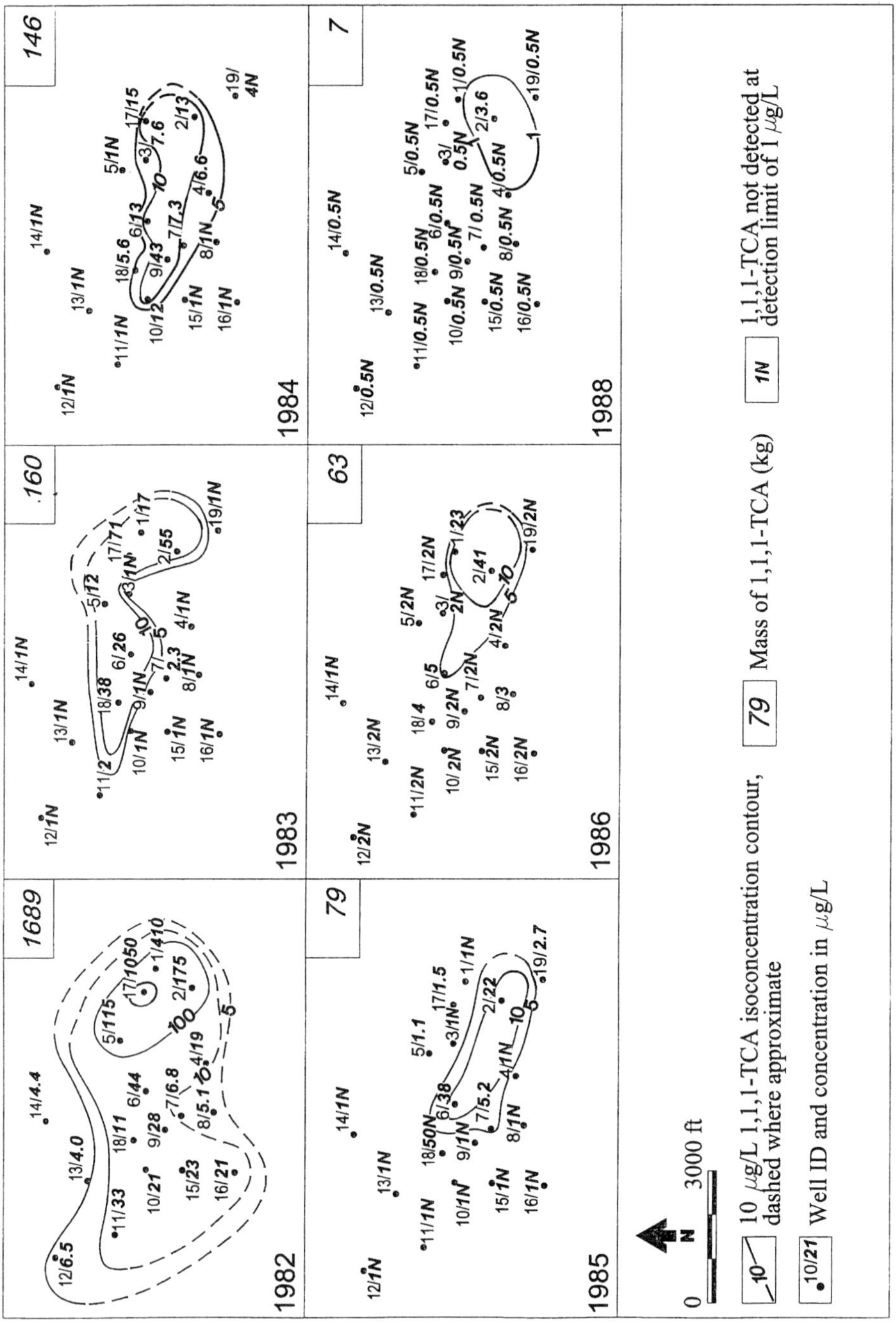

FIGURE 1. Maximum 1,1,1-TCA concentration, 1982-1988

TABLE 2. Maximum 1,1,1-TCA in groundwater (µg/L)

Well #	1982	1983	1984	1985	1986	1988
1	410	17	---	1N	23	0.5N
2	175	55	13	22	41	3.6
3	---	1N	7.6	1N	2N	0.5N
4	19	1N	6.6	1N	2N	0.5N
5	115	12	1N	1.1	2N	0.5N
6	44	26	13	38	5	0.5N
7	6.8	2.3	7.3	5.2	2N	0.5N
8	5.1	1N	1N	1N	3	0.5N
9	28	1N	43	1N	2N	0.5N
10	21	1N	12	1N	2N	0.5N
11	33	2	1N	1N	2N	0.5N
12	6.5	1N	1N	1N	2N	0.5N
13	4.0	1N	1N	1N	2N	0.5N
14	4.4	1N	1N	1N	1N	0.5N
15	23	1N	1N	1N	2N	0.5N
16	21	1N	1N	1N	2N	0.5N
17	1050	1N	15	1.5	2N	0.5N
18	11	38	5.6	50N	4	0.5N
19	---	1N	4N	2.7	2N	0.5N

1N = 1,1,1-TCA not detected at detection limit of 1 µg/L

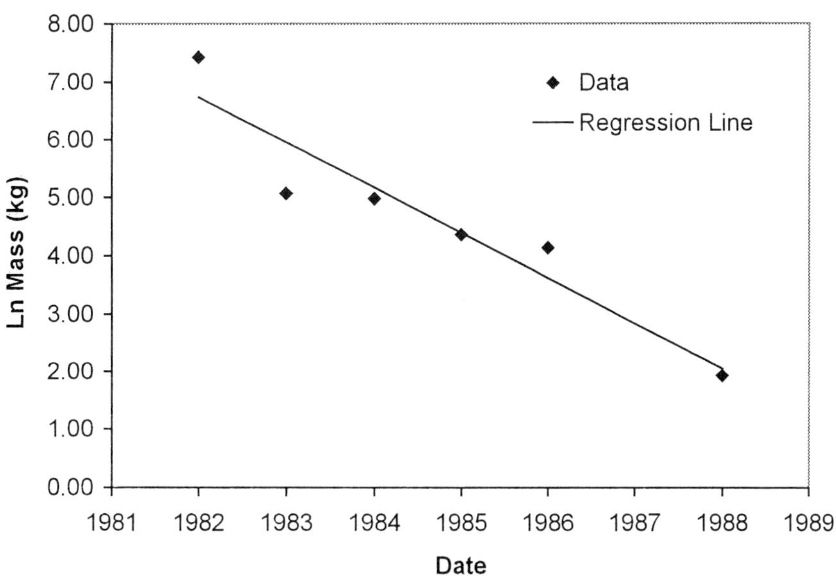

FIGURE 2. Estimated TCA mass vs. time

THE DIOXIN CONUNDRUM: SOURCES, SINKS, AND NATURAL ATTENUATION

Peter Adriaens, Q. Shiang Fu, Iris D. Albrecht, and Andrei L. Barkovskii (The University of Michigan, Ann Arbor, MI, USA)

ABSTRACT: The discrepancy between source, sink, and air-water interface patterns of polychlorinated dibenzo-p-dioxins (PCDD) has been interpreted in light of the reactivity of PCDD in sediments, and microbial or dissolved organic matter (DOM) model systems under estuarine and freshwater conditions. In all systems, the yield of lesser chlorinated products never exceeded 20% of the initial dioxin present and diCDD were the final homologue group produced, except for in sediments where dechlorination proceeded to monoCDD. Biotic and DOM-mediated dechlorination patterns could be distinguished, based on the ratio of 2,3,7,8- to non-2,3,7,8-PCDD (and TCDD), which indicated that DOM strongly influences PCDD dechlorination in sediments. The kinetics of 2,3,7,8-TCDD production from higher chlorinated isomers vs. its further dechlorination to mono- and diCDD indicates that the former will tend to transiently accumulate. The elevated presence of 2,3,7,8-TCDD in most sediments (even those which did not receive a significant input of this isomer) and the accumulation of diCDD at the air-water interface may suggest that dioxins are subject to natural attenuation, ultimately resulting in 'outgassing' in the water column.

INTRODUCTION

Dioxins are ubiquitous trace (ppb) or ultra-trace (ppq) byproducts, predominantly generated from combustion and chlorophenol-based chemical synthesis processes. Urban waterways are affected by atmospheric and direct discharge of polychlorinated dibenzo-*p*-dioxins (PCDD) resulting in a dioxin 'sediment pattern' which is generally characterized by a dominance of hepta- and octaCDD, with minor contributions of hexa- to tetraCDD (Fiedler et al., 1996). Elevated concentrations of the 2,3,7,8-TCDD isomer tend to be associated with direct discharge from sources such as 2,4,5-trichlorophenate production (Bopp et al., 1996). Recently, the accumulation of PCDD at the air-water particulate interface in the Lower Hudson River Esturay was described, exhibiting a profile dominated by octa- and diCDD (Lohmann et al., 2000). Fugacity ratio calculations implied a net volatilization flux from the water body, indicating that sediments may serve as a source of dioxins. Since sediments are not routinely analyzed for mono- to triCDD, it is not clear whether the diCDD are original contaminants, or may have resulted from environmental transformation reactions. Despite the environmental stability of these compounds, a number of literature reports throughout the last decade have indicated that under reducing conditions prevailing in sediments dioxins may undergo dechlorination reactions.

Objective. The overall objectives of this body of research were to (1) construct an inventory of biotic and abiotic dioxin reactivity in sediments, to (2) evaluate methods to stimulate the dechlorination kinetics in sediments based on organic and inorganic electron donors, as well as priming compounds (e.g. 2-monobromoDD), and to (3) interpret the potential for dioxin reactivity in sediments in light of proposed outgassing patterns. Specific emphasis was placed on (i) determining the yield and extent of dechlorination, (ii) assessing dominant dechlorination pathways and kinetics, and (iii) evaluating the fate of 2,3,7,8-substituted residues and the potential accumulation of 2,3,7,8-TCDD.

Site Description.

The last ten years of dioxin research in our laboratory on the capability of sediments and sediment components to mediate dechlorination reactions were predominantly focused on the Passaic River estuary (Fig. 1).

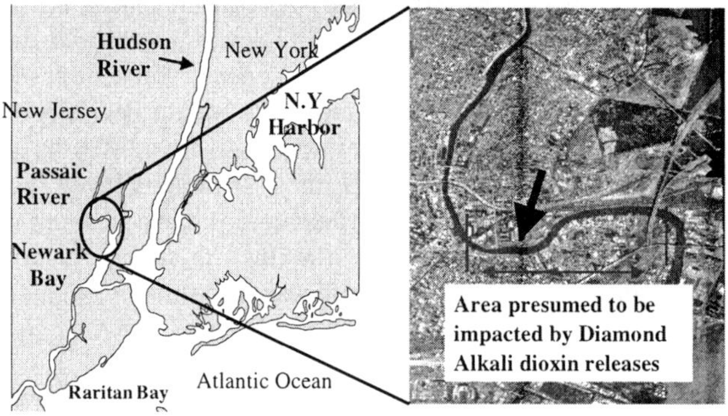

FIGURE 1. Plan view of the Passaic River (NJ) estuary

This waterway has been impacted by atmospheric and direct discharge from chlorophenol production since the 1960s (Bopp et al. 1991), resulting in a dioxin pattern characterized by (Fig. 2): (i) the presence of tetra- to octaCDD/F with a predominance of hepta- and octaCDD/F, (ii) a dominance of 2,3,7,8-TCDD/F within the TCDD/F homologue group, (iii) 1,2,3,7,8-PeCDD dominant within the PeCDD homologue, and (iv) a 2,3,7,8-TCDD/F ratio of approximately 1. The dioxin profiles indicate that concentrations are not log-normal distributed, and show some of the highest recorded 2,3,7,8-TCDD concentrations (350 ng/kg, with peaks in excess of 100 μg/kg). The USEPA maintains a temporal and spatial database with sediment concentrations measured over a 15-year timeframe. Unfortunately, the locations and depths vary, which does not allow for a comparison and assessment of natural attenuation processes.

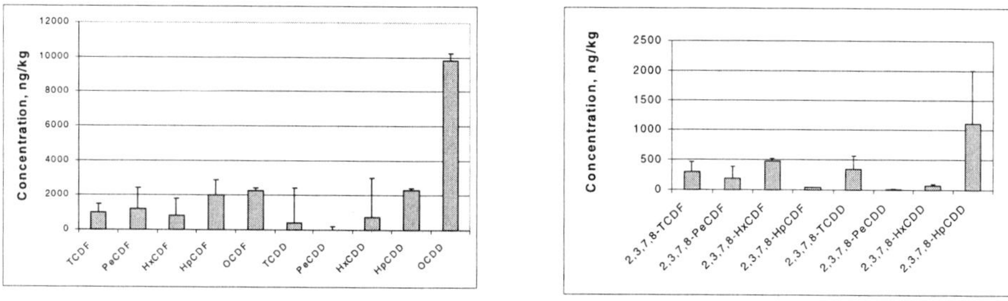

FIGURE 2. Dioxin/furan concentration profiles in the Passaic River

MATERIALS AND METHODS

Over the last decade, two sediment cores were collected near the outfall of 24 Lister Ave. (indicated by the arrow in Figure 1), one of the presumed sources of the dioxin contamination. The cores (9") was cut in 1" sections, sealed with wax, and shipped to The University of Michigan, where they were stored at 4°C until use. Saline (22 parts-per-thousand, ppt) anaerobic river bottom water (core seepage water) was collected at the same time, and was used for all sediment incubations.

The experimental set-ups have been described in detail in our previous publications (Adriaens and Grbic-Galic, 1994; Adriaens et al.,1995; 1996; Barkovskii et al., 1996; 1998; Albrecht and Adriaens, 1997; Albrecht et al., 1999; Fu et al., 1999; 2000), but the generalized experimental matrix is presented in Table 1.

TABLE 1. Experimental matrix of dioxin dechlorination experiments.

Sediments	**Microorganisms**	**Model DOM**
Passaic River cores *Hudson River core*	*Sediment-eluted mixed communities*	*<u>Monomers</u>:* *Catechol, resorcinol, 3,4-dihydroxybenzoic acid* *<u>Polymers</u>:* *Polymaleic acid, Aldrich humic*
<u>Incubation conditions</u>: *Freshwater-spiked Penta- to octaCDD* *Estuarine-historical residues*	*<u>Incubation conditions</u>:* *Freshwater-spiked OCDD* *Freshwater-hist. residues* *Estuarine-spiked HpCDD (both isomers separately)* *Marine-spiked HpCDD*	*<u>Incubation conditions</u>:* *Estuarine-spiked HpCDD (both isomers)*
<u>Electron donors/primers</u>: *Organic acids* *Hydrogen* *2-MonobromoDD*	*<u>Electron donors</u>:* *Organic acids* *(Hydrogen?)*	*<u>Electron donors</u>:* *Sulfide (from reduced anaerobic media)* *Ti-citrate* *Sediment microorganisms*
<u>Electron acceptors</u>: *Bicarbonate,* *Natural (core water)*	*<u>Electron acceptors</u>:* *Bicarbonate* *Sulfate*	*<u>Electron acceptors</u>:* *None*

All incubations were analyzed (isomer-specifically whenever possible) for mono-to octaCDD at two to five timepoints (depending in the incubation condition) to obtain dechlorination (i) yields and homologue distribution, (ii) pathways, and (iii) kinetics.

RESULTS AND DISCUSSION

Dechlorination Yields and Homologue Distributions in Biotic and Abiotic Model Systems. Dioxin dechlorination occurred under all incubation conditions, albeit with significant differences in yield (after 6 month incubations) and distribution of lesser chlorinated products. When the yields of lesser chlorinated PCDD were expressed relative to the disappearance of parent compound, the following observations can be made (Fig. 3): (i) dechlorination does not appear to be a major pathway, as the products account for 5-20%; (ii) the presence of a strong reductant such as Ti-citrate increases the yield 3-4 fold to similar levels observed under high salinity biotic conditions; (iii) the yield is highly dependent on porewater geochemistry. The low mass balance has been explained by the occurrence of other reactions such as transchlorination (migration of chlorine from the dioxin to DOM) and polymerization processes, which have not been quantified to date. Furthermore, it appears that microbial and DOM-mediated reactions are additive, at least based on freshwater incubation conditions.

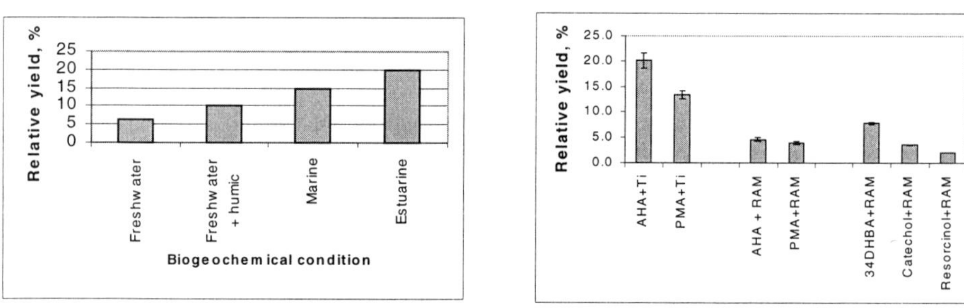

FIGURE 3. Dechlorination yields in biotic and abiotic systems (modified from Adriaens et al., 1999 and Fu et al., 1999)

The homologue group distribution under estuarine conditions after the same incubation period shows that diCDD represent the final extent of dechlorination, at 5-10% for DOM and 10-20% for biotic transformations. Within the latter, 2,3- and 2,7-diCDD were dominant. These distributions are in contrast with freshwater biotic and estuarine sediment incubations where 2-monoCDD accumulated as the final product of dechlorination (Barkovskii and Adriaens, 1996; Albrecht et al., 1999). It is possible that diCDD were not detected in sediments on account of interference of co-extractants.

Dechlorination Pathways in Biotic, Abiotic, and Sediment Systems. Dechlorination pathways can be distinguished based on the production of 2,3,7,8-/non-2,3,7,8-PCDD and –

TCDD ratios. The magnitude of this ratio reflects either a dominance of peri (1,4,6,9)-chlorine removal in case the ratio is >1 (peri-dechlorination pathway), or of lateral (2,3,7,8)-chlorine removal if the ratio < 1 (mixed peri/lateral dechlorination pathway). This is exemplified for TCDD fractionation in Figure 4.

These data indicate that: (i) all ratios are < 1; (ii) the relative fraction of 2,3,7,8-TCDD during biotically-mediated dechlorination is 2-4 fold less than during DOM reactions; (iii) the presence of DOM in microbial incubations shifts the reactions towards peri-dechlorination; (iv) there is an apparent dependence of fractionation on biogeochemical conditions resulting in an increase in ratio with decreasing salinity. This strong fractionation in DOM or DOM+cells incubations is similar to that observed in sediments (Albrecht et al., 1999), where only production of 2,3,7,8-TCDD was observed. Moreover, the fractionation observed for TCDD was representative of that observed for the total PCDD; in that case, the ratio was > 1 for DOM-mediated reactions, and <1 for biotic reactions (Fu et al., 2000).

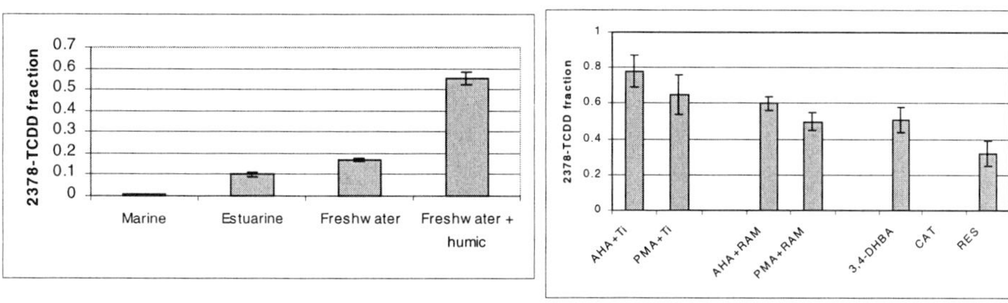

FIGURE 4. TCDD Ratios in biotic and abiotic incubations (modified from Adriaens et al., 1999 and Fu et al., 2000)

Dechlorination Kinetics in Biotic, Abiotic, and Sediment systems. The reaction rates for production of the various homologue groups in the sediment-free model systems generally decrease with decreasing chlorine content, and are on the order of 0.002-0.2 nmol/d. Of particular interest in the overall reactivity of dioxins are the rates of 2,3,7,8-TCDD production vs. its further dechlorination (Table 2).

Whereas production rates can be reasonably easily measured, as often isomer-specific information is available, the dechlorination of 2,3,7,8-TCDD fraught with uncertainty, considering the lack of particularly tri- and diCDD isomers which are commercially available, and the possibility that even identified mono- to triCDD can be derived from TCDD other than the 2,3,7,8-substituted isomer. For example, 2,3,7-triCDD can be formed from 2,3,7,8-TCDD, as well as from 1,2,3,7-TCDD. The only unambiguous case is where 2-monoCDD production was coupled to disappearance of 2,3,7,8-TCDD, as was the case for historical residues which had partitioned to eluted cells (Table 2, first line). With these caveats in mind, Table 2 presents best-case scenarios for production and dechlorination (as evidenced by mono- to triCDD production). Hence, except for in the case of spiked cells under freshwater conditions, and hydrogen-stimulated sediments, the 2,3,7,8-TCDD isomer is likely to accumulate, or at best remain at steady state concentrations during dechlorination reactions starting with the hepta- or octaCDD which are found to dominate in sediments.

TABLE 2. Dioxin dechlorination rates in spiked and historical Passaic River matrices (modified from Albrecht et al., 1999)

Environmental System	*Production Rates*, pmol/d*			
	OCDD	*1,2,3,4,6,7,8-HpCDD*	*2,3,7,8-TCDD*	*2-MCDD (di/triCDD)*
Solid-free systems				
Nonspiked cells#	NA	NA	-0.47±0.29	0.23±0.03
Spiked sediment-eluted cells	-38.8±6.2	1.80±0.13	0.14±0.004	0.33±0.02
Cells + 3,4-DHBA**	-75.3±9.4	2.10±0.17	0.83±0.06	0.43±0.03
3,4-DHBA + RAM	NC	2.10±0.15	1.23±0.04	ND
AHA	NA	-15.3±3.0	3.3±0.2	(14±2)
AHA + Ti-citrate	NA	-131.6±22.0	32.5±5.2	(56±5)
PMA	NA	-6.3±0.4	2.7±0.4	(12±3)
PMA + Ti-citrate	NA	-75.5±3.7	20.4±3.5	(26±6)
Soil/sediment systems				
Autoclaved	0.75	-0.28	0.36	0
Organic acid	-5.66	3.97	0.08	<0.005
Brominated	-5.24	-0.30	2.60	0.06
Hydrogen	-6.76	-8.94	-28.59	79.32

*The rates were based on 25 g DW of sediments, and 5 mL of liquid for the sediment and solid-free systems, respectively. Solid-free systems were spiked with OCDD dissolved in decane to an initial concentration of 11.4 μmoles/L (• 5 mg/L). ** 3,4-DHBA, dihydroxybenzoic acid; # Only 2,3,7,8-TCDD was found to have selectively partitioned to microorganisms in sediment cores; ND, not detected; NA, not applicable; NC, not calculated.

Contributions of Sediment Reactivity to Natural Attenuation of Dioxins. Though natural attenuation of dioxins was not directly measured in this series of studies, it is apparent that under the proper conditions dioxins can undergo extensive transformation reactions, with dechlorination processes representing at best 20% of total reactivity. Several lines of evidence may point towards a role for sediment reactivity as a natural attenuation mechanism. First, the accumulation of diCDDs from biotic and abiotic incubations coincide with profiles observed at the air-water interface which, based on fugacity calculations indicate that the water column and particulates are the source (Lohmann et al., 2000). No isomer-specific analysis has been conducted on these samples to verify whether 2,3- and 2,7-diCDD are present and/or dominant. Second, some core samples showed evidence for the presence of 2-monoCDD (Fig. 5, top panel), but it is not known whether this isomer was originally present or produced from dechlorination (Albrecht and Adriaens, 1997). When other sediment cores, where the original PCDD residues did not include 2-monoCDD (bottom panel), were incubated in the presence of hydrogen gas (treatment 5), the profile changed to show a decrease in octa- and tetraCDD, and production of 2-monoCDD (25 mol%). The microbial capability to dechlorinate historical 2,3,7,8-TCDD residues to 2-monoCDD was also observed earlier (Barkovskii and Adriaens, 1996). Extrapolation of the rates at which these reactions are likely to occur are difficult, as some assumptions need to be made with respect to dominant catalysts of dechlorination activity in sediments, bioavailability of the contaminant, and electron fluxes in reduced sediments.

Assuming the reaction is cometabolic (ratio ≈ 1:1000 chlorine removal to methane/sulfate generation), and is coupled to either methanogenesis or sulfate-reduction, the sediment reaction rate is on the order of 5-10 years per chlorine removed, based on in situ measured respiration rates.

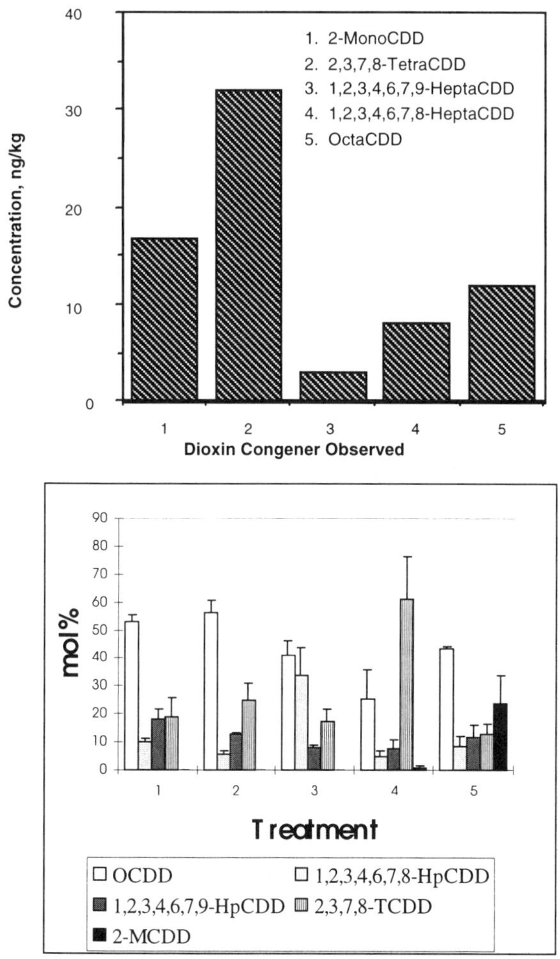

FIGURE 5. Dioxin residue profiles in historical cores (top panel), and after amendments (bottom panel) (legend: 1. original profile; 2. autoclaved; 3. organic acid amendment; 4. 2-monobromoDD amendment; 5. hydrogen amendment) (modified from Albrecht et al., 1999).

ACKNOWLEDGEMENTS

The authors acknowledge support from the US Environmental Protection Agency, the EPA Region II Superfund Office, the Army Corps of Engineers-Waterways Experiment Station, and the Office for Naval Research for funding this work between 1990-2000. Since this manuscript has not been reviewed by these agencies, no endorsement should be inferred.

REFERENCES

Adriaens, P. and D. Grbic'-Galic'. 1994. "Reductive Dechlorination of PCDD/F by Anaerobic Cultures and Sediments." *Chemosphere* 29: 2253-2259.

Adriaens, P., Q. Fu, and D. Grbic'-Galic. 1995. "Bioavailability and transformation of highly chlorinated dibenzo-p-dioxins and dibenzofurans in anaerobic soils and sediments." *Environ. Sci. Technol.* 29: 2252-2261.

Adriaens, P., P.R.-L. Chang, and A.L. Barkovskii. 1996. " Dechlorination of Chlorinated PCDD/F by Organic and Inorganic Electron Transfer Molecules in Reduced Environments". *Chemosphere* 32: 433-441.

Albrecht, I.D., and P. Adriaens. 1997. "Evaluating the Fate of Dioxins in Reduced Sediments: Evidence for Natural Reductive Dechlorination Processes." In H. Verachtert and W. Verstraete (Eds.), *International Symposium for Environmental Biotechnology (ISEB),* pp. 405-408. Technological Institute, Antwerp, Belgium.

Albrecht, I.D., A.L. Barkovskii, and P. Adriaens. 1999. "Production and Dechlorination of 2,3,7,8-Tetrachloro-Dibenzo-p-Dioxin (TCDD) in Historically-Contaminated Estuarine Sediments." *Environ. Sci. Technol.*33: 737-744.

Barkovskii, A.L., and P. Adriaens. 1996. "Microbial Dechlorination of Historically Present and Freshly Spiked Chlorinated Dioxins and the Diversity of Dioxin-Dechlorinating Populations." *Appl. Environ. Microbiol.* 62: 4556-4562.

Barkovskii, A.L., and P. Adriaens. 1998. " Impact of Model Humic Constituents on Microbial Reductive Dechlorination of Polychlorinated Dibenzo-p-Dioxins." *Environ. Toxicol. Chem.* 17: 1013-1021.

Bopp, R.F., M.L. Gross, H. Tong, H.J. Simpson, S.J. Monson, B.L. Deck, and F.C. Moser. 1991. "A Major Incident of Dioxin Contamination: Sediments of New Jersey Estuaries." *Environ. Sci. Technol.* 25: 951-956.

Fiedler, H., C. Lau, L.O. Kjeller, and C. Rappe. 1996. Investigations of Source Patterns in the Lower Mississippi River. *Chemosphere* 32: 421-432.

Fu, Q.S., A.L. Barkovskii, and P. Adriaens. 2000. "Dechlorination of peri-Chlorinated Dioxins by Model Humic Acids." *Environ. Sci. Technol.* In Press.

Fu, Q., A.L Barkovskii, and P. Adriaens. 1999. " Reductive Transformation of Dioxins: An Assessment of the Contribution of Dissolved Organic Matter to Dechlorination Reactions." *Environ. Sci. Technol.*33, 3837-3842.

Lohmann, R., Nelson, E., Eisenreich, and K.C. Jones. 2000. "Evidence for Dynamic Air-Water Exchange of PCDD/Fs: A Study in the Raritan Bay/Hudson River Estuary." *Environ. Sci. Technol.* In Press.

NATURAL ATTENUATION OF 1,2-DICHLOROETHANE BY AQUIFER MICROORGANISMS UNDER Mn(IV)-REDUCING CONDITIONS

Paul M. Bradley, Richard S. Dinicola, and James E. Landmeyer
U.S. Geological Survey, Columbia, SC, USA

ABSTRACT: A laboratory investigation was initiated to evaluate the potential for anaerobic, microbial mineralization of 1,2-DCA to CO_2 under Mn(IV)-reducing conditions. Efficient anaerobic oxidation of [1,2-^{14}C] dichloroethane (DCA) to $^{14}CO_2$ under Mn(IV)-reducing conditions was observed. Approximately 65% mineralization of [1,2-^{14}C] DCA was observed after 62 days. No products other than $^{14}CO_2$ were observed under these conditions. The results of this study indicate that anaerobic microbial oxidation under Mn(IV)-reducing conditions may be a significant mechanism for natural attenuation of chloroethane contaminants in ground-water systems.

INTRODUCTION

Chloroethene and chloroethane compounds are prevalent and among the the most toxic contaminants found in anaerobic groundwaters throughout the U.S. (McCarty and Semprini, 1994). Recent investigations have demonstrated that, in addition to reductive dechlorination processes, some chloroethene contaminants can be biodegraded to CO_2 under anaerobic conditions without accumulation of toxic intermediates (Bradley and Chapelle, 1998; Bradley et al., 1998a; 1998b). Thermodynamic considerations indicate that anaerobic microbial oxidation to CO_2 also is favorable for some chloroethanes and suggest that oxidative degradation must be considered when assessing the potential for microbial degradation of chloroethanes in groundwater systems. The exclusive purpose of this study was to test the hypothesis that microbial oxidation can be a significant mechanism for anaerobic biodegradation of chloroethane contaminants.

In this study, the ability of aquifer microorganisms to oxidize 1,2-dichloroethane to CO_2 under Mn(IV)-reducing conditions was examined. 1,2-dichloroethane (1,2-DCA) is a common ground-water contaminant, due to its high aqueous solubility and widespread uses in PVC manufacturing and degreasing activities. 1,2-DCA is listed by the EPA's national Toxic Release Inventory as a carcinogen and has a drinking water MCL of 5 μg/L. At present, anaerobic microbial degradation of 1,2-DCA has been shown to occur only via reductive dechlorination or dichloroelimination (Belay and Daniels, 1987; Egli et al., 1987; Holliger et al., 1990; Klecka et al., 1998). However, thermodynamic considerations (see equation) and a recent report of microbial mineralization of 1,2-dichloroethene (1,2-DCE) to

$$C_2H_4Cl_{2\,(g)} + 2MnO_{2\,(s)} + 2H_2O_{\,(l)} \rightarrow 2HCO_3^-{}_{(aq)} + 2Mn^{++}{}_{(aq)} + 2Cl^-{}_{(aq)} + 3H_{2\,(aq)}$$
$$\Delta G^o_r = -362.4 \text{ kJ/mole}$$

CO_2 under Mn(IV)-reducing conditions (Bradley et al., 1998b) indicate that microbial oxidation of dichloroethane may be significant in Mn(IV)-reducing aquifer systems.

MATERIALS AND METHODS

To test this hypothesis, a microcosm study was conducted using aquifer sediments from a chlorinated solvent contaminated site at the Naval Air Station (NAS) Whidbey Island, Washington. The primary ground-water contaminants at the site were chloroethene and chloroethane solvents. At the point of sediment sample collection, dissolved concentrations of dichloroethane (1,1-DCA and 1,2-DCA) were approximately 100 μg/L. Aquifer sediments at the site were characterized by medium- to fine-grained sands, a Mn(IV) oxide content of 0.05 to 0.1% (dry weight), and dissolved Mn(II) concentrations up to 7.1 mg/L. Field geochemical data indicated and subsequent microcosm incubations confirmed that Mn(IV)-reduction predominated in these sediments under anaerobic conditions. Microcosm preparation, incubation conditions, and analytical procedures have been described previously (Bradley et al., 1998b; Bradley and Chapelle, 1999a; 1999b; Lovley and Phillips, 1988a; 1988b). Sediment microcosms were pre-incubated for 5 days and then amended with approximately 0.05 μCi of [1,2-^{14}C] DCA (specific activity = 11.1 mCi/mmole, radiochemical purity $\geq$98%). Initial dissolved 1,2-DCA concentrations in equilibrium with the headspace were estimated, based on experimentally determined adsorption and Henry's Law coefficients, to be 100 μg/L.

RESULTS AND DISCUSSION

Microorganisms indigenous to the aquifer at NAS Whidbey Island were capable of significant mineralization of [1,2-^{14}C] DCA to $^{14}CO_2$ under unamended anaerobic conditions (Figure 1). Approximately 64% mineralization of [1,2-^{14}C] DCA was observed after 62 days. DCA mineralization was attributable to biological activity because the final recovery of $^{14}CO_2$ in killed control microcosms was less than 2%. The lack of dissolved O_2 ([O_2] < 3 μM), NO_3 ([NO_3] < 0.2 μM), and SO_4 ([SO_4] < 20 μM), the lack of production of CH_4 (not detected, [CH_4] < 1 μmole/liter headspace), dissolved sulfide (not detected, [HS] < 0.2 μM), and dissolved Fe(II) ([Fe(II)] < 0.2 μM), and the observed accumulation of Mn(II) (8±3 nmoles produced, Table 1) within these experimental microcosms indicated that Mn(IV)-reduction was the predominant terminal electron accepting process under unamended, anaerobic conditions. These results demonstrate that chloroethane contaminants are susceptible to oxidative microbial degradation under anaerobic conditions.

Addition of poorly crystalline MnO_2 stimulated overall microbial respiration but did not significantly effect [1,2-^{14}C] DCA mineralization in this study (Figure 1). The effect of MnO_2 treatment on microbial respiration in general was evaluated by analyzing the change in total headspace CO_2 concentration throughout the incubation (Figure 1). Amendment of anaerobic microcosms with MnO_2 resulted in a four-fold

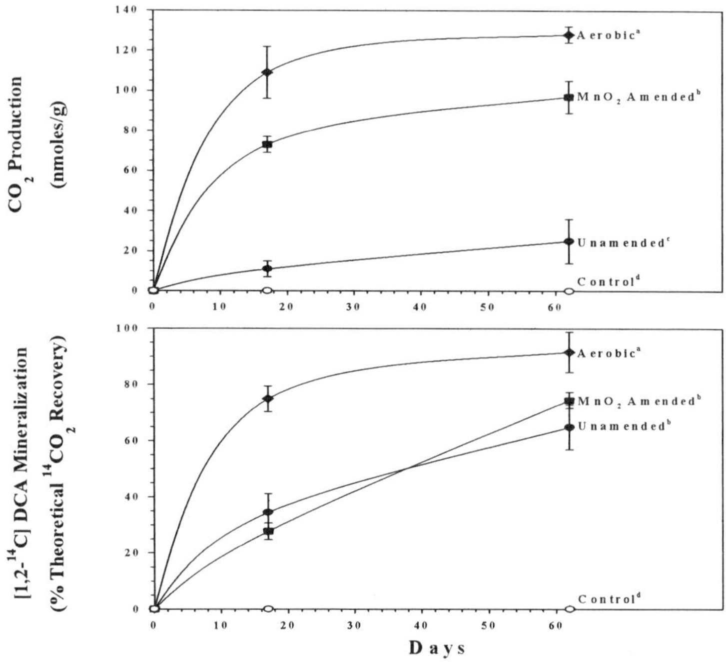

Figure 1. Total CO_2 production (nmoles/g sediment dry weight) and percentage mineralization of [1,2-^{14}C] DCA to $^{14}CO_2$ in microcosms containing aquifer material from NAS Whidbey Island. Experimental data are means$\pm$SD for triplicate microcosms and the control data are from duplicate sterile control microcosms. The final mean $^{14}CO_2$ recovery of treatments with different superscripts was statistically different according to the Kruskal-Wallis one-way ANOVA on ranks ($p<0.05$).

increase in total CO_2. Likewise, MnO_2 amendment resulted in an apparent doubling of Mn(IV)-reduction (Table 1). These results indicate that microbial respiration and Mn(IV)-reduction were limited by the bioavailability of Mn(IV) in these aquifer sediments. The final recovery of [1,2-^{14}C] DCA radioactivity as $^{14}CO_2$ was slightly higher in MnO_2 amended microcosms than in unamended anaerobic microcosms, but the difference was not statistically significant (Figure 1, Table 1). This observation indicates that Mn(IV) bioavailability was not a primary limitation on microbial oxidation of [1,2-^{14}C] DCA under these study conditions (Figure 1).

No volatile intermediates were detected in this study under anaerobic conditions. Because chloroethane, ethane and ethene were previously reported as products of anaerobic microbial degradation of 1,2-DCA under SO_4-reducing and methanogenic conditions (Belay and Daniels, 1987; Egli et al., 1987; Holliger et al., 1990; Klecka et al., 1998), the headspace gas of aquifer microcosms was analyzed for the presence of ^{14}C-chloroethane, ^{14}C-ethane and ^{14}C-ethene throughout the 62 day

Table 1. Production of dissolved Mn(IV) and the final recovery of ^{14}C as $^{14}CO_2$ in aquifer microcosms.[a]

Microcosm	nmoles Mn(II) produced[b]		% recovery as $^{14}CO_2$	
	Exptl	Control	Exptl	Control
Anaerobic	8±3	NS	64±8	NS
MnO_2 amended	20±3	NS	75±3	NS
Aerobic	NS	NS	92±7	NS

[a]62-day incubation. Data are means±standard deviations for triplicate experimental microcosms and duplicate control microcosms.
[b]The amount of Mn(II) produced is presented as a qualitative indicator of Mn(IV)-reduction, because the Mn(II) found in the dissolved phase is a small fraction (less than 10%) of the total Mn(II) present in the system. NS, no significant change.

incubation period. None of these compounds was detected in this study (detection limit < 5 nM). After 62 days, the amount of [1,2-^{14}C] DCA remaining in the microcosms was determined by radiometric headspace analysis. No [1,2-^{14}C] DCA was detected in any experimental treatment (detection limit < 10 nM or about 10% of the initial, [1,2-^{14}C] DCA headspace concentration), but headspace concentrations (corrected for sample removal) in control microcosms did not change significantly over the study period (93±4% final recovery). The absence of detectable, volatile intermediates in this study suggests that the 20-30 % of the radioactivity that was not recovered under anaerobic conditions was incorporated into biomass or present as non-volatile products.

This investigation is the first report of anaerobic oxidation of 1,2-DCA to CO_2 under Mn(IV)-reducing conditions. Mn(IV) oxides are potentially powerful oxidants which are common in aquifer sediments and ground-water systems. Thus these results further illustrate the potential importance of Mn(IV) reduction for remediation of chlorinated ground-water contaminants. Most important, the results of this study indicate that chloroethane contaminants, like chloroethene contaminants, are susceptible to microbial oxidation under anaerobic conditions.

ACKNOWLEDGMENTS

This study was supported by Sonia Murphy of the Engineering Field Activity, Northwest, Department of the Navy and the U.S. Geological Survey Toxic Substances Hydrology Program.

REFERENCES

Belay, N., and L. Daniels. 1987. "Production of ethane, ethylene, and acetylene from halogenated hydrocarbons by methanogenic bacteria." *Appl. Environ. Microbiol.* 53: 1604-1610.

Bradley, P. M., and F. H. Chapelle. 1998. "Microbial mineralization of VC and DCE under different terminal electron accepting conditions." *Anaerobe* 4: 81-87.

Bradley, P. M.; F. H. Chapelle, and D. R. Lovley. 1998. "Humic acids as electron acceptors for anaerobic microbial oxidation of vinyl chloride and dichloroethene." *Appl. Environ. Microbiol.* 64: 3102-3105.

Bradley, P. M., J. E. Landmeyer, and R. S. Dinicola. 1998. "Anaerobic oxidation of [1,2-^{14}C] dichloroethene under Mn(IV)-reducing conditions." *Appl. Environ. Microbiol.* 64: 1560-1562.

Bradley, P. M., and F. H. Chapelle. 1999. "Methane as a product of chloroethene biodegradation under methanogenic conditions." *Environ. Sci. Technol.* 33: 653-656.

Bradley, P. M., and F. H. Chapelle. 1999. "Role for acetotrophic methanogens in methanogenic biodegradation of vinyl chloride." *Environ. Sci. Technol.* 33: 3473-3476.

Egli, C., R. Scholtz, A. M. Cook, and T. Leisinger. 1987. "Anaerobic dechlorination of tetrachlomethane and 1,2-dichloroethane to degradable products by pure cultures of *Desulfobacterium* sp. and *Methanobacterium* sp." *FEMS Microbiol. Lett.* 43: 257-261.

Holliger, C., G. Schraa, A. J. M. Stams, and A. J. B. Zehnder. 1990. "Reductive dechlorination of 1,2-dichloroethane and chloroethane by cell suspensions of methanogenic bacteria." *Biodeg.* 1: 253-261.

Klečka, G. M., C. L. Carpenter, and S. J. Gonsior. 1998. "Biological transformations of 1,2-dichloroethane in subsurface soils and groundwater." *J. Contam. Hydrol.* 34: 139-154.

Lovely, D. R. and E. J. P. Phillips. 1988. "Manganese inhibition of microbial iron reduction in anaerobic sediments." *Geomicrobiol.* J. 6: 145-155.

Lovely, D. R. and E. J. P. Phillips. 1988. "Novel mode of microbial energy metabolism: organic carbon oxidation coupled to dissimilatory reduction of iron and manganese." *Appl. Environ. Microbiol.* 54: 1472-1480.

McCarty, P. L., and L. Semprini. 1994. "Ground-water treatment for chlorinated solvents." p. 87-116. *In* R. D. Norris et al. (eds.) *Handbook of Bioremediation*, Lewis Publishers, Boca Raton.

NATURAL ATTENUATION OF CHLORINATED BENZENES AT A FORMER DISPOSAL SITE

R. Todd Herrington, ***John Hicks***, and Douglas Downey (Parsons Engineering Science, Inc., Denver, Colorado)
Jim Spain and Shirley Nishino (Air Force Research Laboratory, Tyndall Air Force Base, Florida)
Erica Becvar (Applied Research Associates, Inc., Panama City, Florida)
Jim Gossett (Cornell University, Ithaca, New York)

Abstract: This case study examines a former military waste disposal site at Kelly Air Force Base (AFB), Texas, where the surficial aquifer is contaminated with dichlorobenzenes (DCBs) and chlorobenzene (CB) from past degreasing activities. An investigation was completed to assess whether DCBs and CB were being attenuated via biodegradation, and to evaluate the implementability of monitored natural attenuation (MNA) as an alternative in ongoing feasibility studies and corrective measure studies for the site. This paper reviews the biochemistry and microbiology of chlorinated benzene biodegradation, the results of microcosm studies for the site, current and historic chemical concentration trends, and geochemical trends, biodegradation rates, and modeling results. The areal extent of the current groundwater plume is stabilized due to the combined effects of aerobic and anaerobic biodegradation. Evidence suggests that DCB is reductively dehalogenated to CB, while CB is removed from groundwater by aerobic biodegradation at the plume perimeter. MNA will provide protection of potential downgradient receptors, although source treatment would decrease the length of time required for long-term monitoring.

INTRODUCTION

Natural physical, chemical, and biological processes often provide beneficial stabilization and destruction of groundwater contaminants. The measurement and utilization of these natural processes in a managed remedial approach is referred to as MNA. Chlorinated benzenes in groundwater appear to be well-suited to the application of MNA because they are selectively biodegradable under aerobic and anaerobic conditions, and exhibit biodegradation behaviors similar to those of chlorinated ethenes, which have been the subject of much study. This paper describes the natural attenuation of a mixed DCB and CB groundwater plume at Site S-1, Kelly AFB, San Antonio, Texas. The work was performed as part of an ongoing study sponsored by the United States Air Force Research Laboratory to conduct site characterization and groundwater modeling to evaluate the scientific defensibility of MNA as a remedial option for groundwater contaminated with chlorinated benzenes (Parsons Engineering Science, Inc., 1999).

CHLORINATED BENZENE BIODEGRADATION

Chlorinated benzene biodegradation has been documented in the laboratory, but little information is available regarding biodegradation of these compounds in the environment. CB and polychlorinated benzenes up to and including tetrachlorobenzene (TeCB) have been shown to be biodegradable under aerobic conditions (Spain, 1997). Nishino *et al.* (1994) note that aerobic bacteria able to metabolize CB have been detected at a variety of CB-contaminated sites, but not at uncontaminated sites. Spain (1996) notes that this provides strong evidence that the bacteria are selected for their ability to adapt and to derive carbon and energy from *in situ* chlorinated benzene biodegradation.

For the highly chlorinated benzenes (e.g., hexachlorobenzene and pentachlorobenzene, as well as TeCB, and trichlorobenzene [TCB]), reductive dechlorination in anaerobic environments is the most likely biodegradation mechanism (Holliger *et al.*, 1992; Ramanand *et al.*, 1993; Suflita and Townsend, 1995). The chlorinated benzenes most likely act as electron acceptors as other sources of carbon and energy are being utilized by microbes or microbial consortia (Suflita and Townsend, 1995). As these compounds are dechlorinated, the daughter products become more resistant to further reductive dechlorination (Fathepure *et al.*, 1988; Bosma *et al.*, 1988; Holliger *et al.*, 1992), but are increasingly amenable to aerobic biodegradation (Schraa *et al.,* 1986; Spain and Nishino, 1987; Ramanand *et al.*, 1993). The reductive dechlorination of CB has not been documented; however, the structurally similar compound chlorotoluene has been observed to be dehalogenated to toluene in methanogenic soil slurries (Ramanand *et al.*, 1993).

Numerous studies have documented the existence of chlorinated benzene-degrading bacteria at contaminated sites. However, studies suggest that indigenous bacteria in uncontaminated ecosystems typically are not capable of metabolizing chlorinated benzenes, but become capable through the exchange and incorporation of genetic information between two separate bacterial types (van der Meer *et al.*, 1998). Indigenous microorganisms aerobically degrade CB and DCBs in groundwater after an acclimation period. The length of the acclimation period may determine when and where biodegradation begins at a CB spill site. Selective enrichments from soil and water samples have taken from 6 to 14 months to develop the ability to degrade 1,3-DCB, 1,4-DCB, 1,2-DCB, and 1,2,4-TCB (de Bont *et al*, 1986; Schraa *et al.*, 1986; Spain and Nishino, 1987; Haigler *et al.*, 1988; and van der Meer *et al.*, 1987).

SITE S-1 DESCRIPTION

Kelly AFB is located in south-central Texas, 7 miles southwest of the center of San Antonio. Site S-1 was used as an intermediate storage area for mixed solvents, carbon cleaning compounds, and waste petroleum, oils, and lubricants from the 1960s until 1973 (Figure 1). Wastes were stored in aboveground tanks that were located in an open pit, or sump area. Turco Carbo Strip was used extensively by the Air Force for removal of carbon from jet engine parts. This solvent was composed of 1,2-DCB and 25-percent cresylic acid. It is probable that Kelly AFB used similar solvent formations for various industrial operations, though a history of the sources of CB and DCB is unavailable for Site S-1.

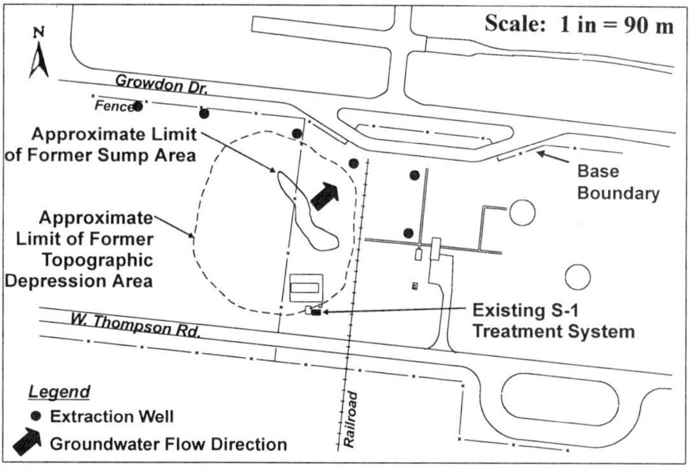

FIGURE 1. Layout of Site S-1, Kelly AFB, TX.

Geology and Hydrogeology. The shallow aquifer overlying the Navarro Clay aquitard is composed of coarse-grained alluvial sand and clayey gravel. The saturated thickness of the alluvial aquifer within the source area in March/April 1998 ranged from approximately 1 to 5 feet (0.3 to 1.5 m). The hydraulic conductivity of the alluvial sands ranges from 7.1 x 10^{-3} cm/sec to 6.7 x 10^{-1} cm/sec. The groundwater velocity in the former sump area was estimated to be 4.6 ft/day (1.4 m/day), decreasing to approximately 0.5 ft/day (0.2 m/day) to the northeast (downgradient). An interim pump-and-treat system installed to prevent off-Base migration influences groundwater flow directions and rates; however, the thinness of the saturated zone limits the radius of influence of this system.

Contaminant Source. Free and residual product within an area of approximately 2.2 acres (9,000 m^2) constitutes the source of chlorinated benzene contamination at Site S-1. At the center of the former sump area, large volumes of product are present. A fluctuating water table and the relatively high groundwater velocities have caused vertical smearing and downgradient migration of free product from the sump. The average thickness of the smear zone along the water table is 7 to 10 feet (2 to 3 m). Free product primarily consists of CB and DCBs with smaller fractions of fuel hydrocarbons.

FIELD INVESTIGATION AND AEROBIC MICROCOSM SETUP

Groundwater samples were collected from 25 monitoring wells to assess trends in natural chemical attenuation over time. In addition to the site contaminants, all groundwater samples were analyzed for a suite of natural attenuation indicator parameters.

Aquifer solids and groundwater also were collected from the downgradient portion of the plume (CB concentration of approximately 45 μg/L).

These materials were combined in a column to create an aerobic microcosm matrix. Filter-sterilized groundwater was sparged with oxygen, and the CB content was adjusted to 7.3 mg/L to serve as the feed reservoir. The influent was delivered to the bottom of the column and exited at the top of the column.

NATURE AND EXTENT OF CHLORINATED BENZENE CONTAMINATION IN GROUNDWATER

Of the three DCB isomers detected at Site S-1, 1,2-DCB was present at the highest concentrations in both free product and groundwater. The maximum 1,2-DCB concentration of 10,000 µg/L was detected in the anaerobic source area (Figure 2). The rapid disappearance of all DCB isomers downgradient from the source area, and the presence of elevated CB concentrations, suggest that DCB is being reductively dechlorinated to CB. The DCB plume is approximately 140 feet (43 m) wide by 200 feet (61 m) long. The amount of CB attributable to reductive dechlorination of DCB cannot be determined because CB concentrations also are elevated in free product samples, suggesting that CB also was a source contaminant disposed of at Site S-1.

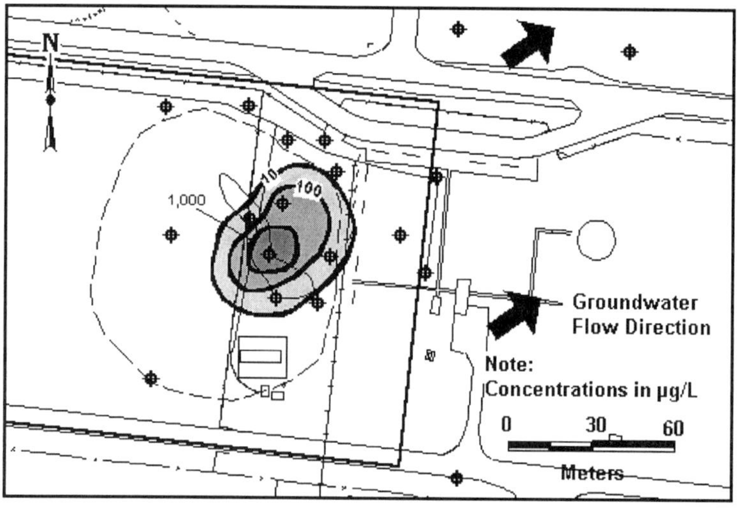

FIGURE 2. 1,2-DCB plume in groundwater, March/April 1998.

The areal extent of DCB in groundwater is relatively small compared to that of CB (Figure 3). The CB groundwater plume emanates from the former sump area and extends 2,000 feet (610 m) to the northeast. The groundwater extraction wells do not appear to be entirely preventing the downgradient migration of dissolved CB. The maximum CB concentration of 45,000 µg/L was detected within the former sump area. The groundwater CB plume at Site S-1 appears to be approximately 1,000 feet (305 m) shorter than was observed a decade ago, indicating a receding plume. Substantial reductions in CB concentrations at downgradient wells between 1989 and 1995, prior to the installation of the groundwater extraction system, indicate that source weathering and natural attenuation of dissolved CB were occurring during this period.

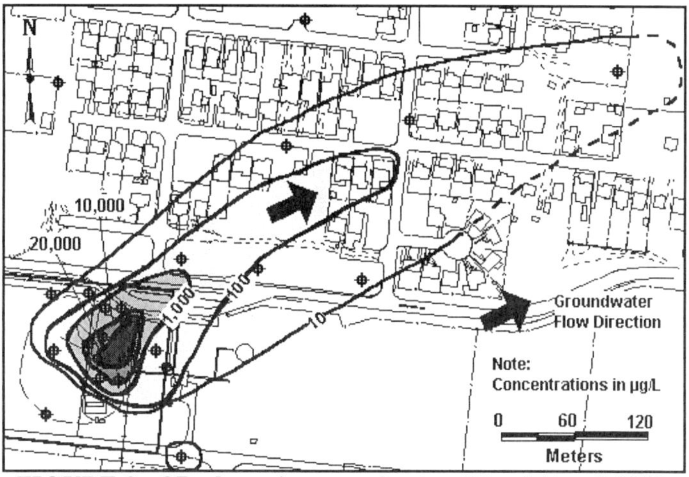

FIGURE 3. CB plume in groundwater, March/April 1998.

Although CB contamination has been present at Site S-1 for at least three decades, it is possible that significant biodegradation did not begin to destroy contaminant mass until the 1980s. Microbial acclimation to CB contamination through the process of genetic recombination requires an extended period of time. Such acclimation may have taken several decades at Site S-1, although historical data are insufficient to support this observation.

GROUNDWATER GEOCHEMISTRY

Groundwater geochemical data support the presence of anaerobic, reducing conditions in the source area and throughout most of the plume. Dissolved oxygen (DO) concentrations ranged from 0.1 to 7.0 mg/L, with the highest concentrations measured at wells crossgradient from the plume. Throughout the plume, DO concentrations generally were less than 0.5 mg/L. The low oxygen concentrations in the plume area relative to uncontaminated areas are a strong indication that microbial activity stimulated by the presence of anthropogenic hydrocarbons (e.g., CB, DCBs, benzene, and other fuel constituents) is depleting DO and creating reducing conditions conducive to DCB dehalogenation.

The area of depleted DO concentrations also is characterized by low oxidation/reduction potentials (ORPs); depleted nitrate and sulfate concentrations; elevated chloride, ferrous iron, methane, and carbon dioxide concentrations; and elevated alkalinity. The observed depletion of electron acceptors and production of metabolic byproducts supports the microbial consumption of anthropogenic organic carbon (e.g., fuel hydrocarbons) in the plume area and indicates that conditions are favorable for reductive dehalogenation of DCBs, especially in the source area.

BIODEGRADATION RATE ESTIMATES

Methods described by Buscheck and Alcantar (B&A) (1995) and Moutoux *et al.* (1996) were used to estimate first-order, field-biodegradation rate constants for chlorinated benzenes at Site S-1. The total destructive attenuation rates estimated using the B&A method for CB downgradient from the source area ranged from 7.9×10^{-4} to 2.2×10^{-3} day^{-1}. Total destructive attenuation rates for CB in the source area ranged from 1.4×10^{-2} to 1.9×10^{-2} day^{-1}. Decay rates computed for DCB in and immediately downgradient from the source area using the Moutoux *et al.* (1996) method indicated a reductive dechlorination rate of 4.6×10^{-3} day^{-1}, compared to a B&A rate estimate of 6.7×10^{-2} day^{-1}.

AEROBIC MICROCOSM STUDY RESULTS AND DISCUSSION

The aerobic microcosm results indicated that viable CB-degrading bacteria were present in the subsurface materials, and that an acclimation period of 3 days was required to establish substantial CB biodegradation rates. Based on the short acclimation period, CB degradation appeared to be caused by induction of an existing CB-degrading microbial population, rather than through a new genetic recombination. A significant microbial population that can acclimate rapidly to degrade CB appears to exist at the fringes of the plume. The results listed in Tables 1 and 2 indicate a substantial capacity for CB degradation, and suggest that the rate might be limited only by the mass transfer of CB and the availability of oxygen.

TABLE 1. Chlorobenzene in various pumped volumes of the soil column operated in intermittent mode.

Days of Operation	Chlorobenzene in Feed Water (mg/L)	Chlorobenzene at 20-mL Pumped Effluent Volume (mg/L)	Chlorobenzene at 120-mL Pumped Effluent Volume (mg/L)
0	7.3	7.9	6.9
1	7.3	6.2	6.5
3	7.4	3.1	5.6
5	5.8	1.7	4.7
6	5.5	1.2	5.0
7	5.6	0.7	4.7

TABLE 2. Chlorobenzene in soil column effluent during continuous feeding.

Flow rate (mL/hour)	Hours of Operation at Specified Flow Rate	Chlorobenzene in Effluent (mg/L)[a/]
35	0.5	0.7
35	14.5	0.0
40	1	0.0
75	1	0.0
130	1	1.6
130	1.5	1.2
200	2.0	2.0
200	2.5	1.1

[a/] Chlorobenzene in influent = 5.5 mg/L.

Acclimated CB-degrading bacteria were detected in significant population numbers at Site S-1 during a previous study (Nishino *et al.*, 1994). Onsite bioreactors installed at Site S-1 by Nishino *et al.* (1994) removed most of the aromatic compounds from the influent groundwater within a 40-minute hydraulic residence time. After a 7-day acclimation period, the federal drinking water standard for CB of 100 μg/L was achieved in the reactor at hydraulic residence times as short as 30 minutes. This confirms that a viable CB-degrading microbial population had pre-existed, or preacclimated, to CB at the site. No changes in CB degradation rates were observed when supplies of inorganic nutrients were discontinued to the reactors, indicating that site groundwater had sufficient nutrients to maintain rapid aerobic biodegradation rates of CB and other compounds in the reactors. When a foreign CB-degrading strain of bacteria (JS150) was added to the reactors, these bacteria did not survive, but were displaced by more viable populations of indigenous Site S-1 bacteria. This result indicated that artificial inoculation of other known strains of CB-degrading bacteria would not be beneficial relative to the natural, preestablished Site S-1 microbial community.

GROUNDWATER MODELING

A numerical groundwater model was developed to predict future plume trends under the influence of natural attenuation and other planned source-reduction actions. The model code Bioplume III (USEPA, 1998) was used to estimate the future fate and transport of dissolved CB under the influence of advection, dispersion, sorption, and biodegradation.

A focused feasibility study performed for Site S-1 (CH2M Hill, 1997) identified limited source excavation and subsequent soil vapor extraction as the preferred remedy to reduce leaching of chlorinated benzenes into the groundwater. Therefore, a source-reduction scenario was simulated that considered the effects of these activities on the longevity of the groundwater plume. The simulation assumed that the source product would be reduced by at least 90 percent during a 4-year treatment period beginning in 1999, followed by natural weathering of the remaining source at a potentially conservative geometric rate of 3 percent per year. The interim groundwater pump-and-treat system was assumed to cease operation in 2000.

By 2004, dissolved CB concentrations were predicted to be one to two orders of magnitude below the 1999 concentrations at most locations (Figure 4), and the CB plume was predicted to recede toward the source area. The model predicted that 36 years after initiation of source-reduction activities would be required before CB concentrations at all off-base wells are below the Federal Drinking Water Standard of 100 μg/L. If source remediation is not conducted at Site S-1, the estimated time for the attainment of this regulatory level at all off-base wells is anticipated to exceed model predictions by decades. Overall, the groundwater model predicted that receptor exposure pathways involving shallow groundwater are not complete under current conditions, nor will the CB plume expand to impact potential downgradient receptors under worst-case (no engineered source reduction) conditions.

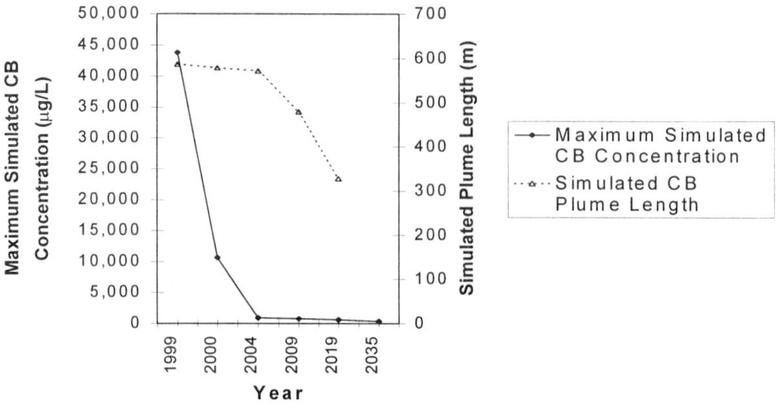

FIGURE 4. Simulated CB plume length and maximum CB concentrations.

DISCUSSION

Aerobic biodegradation of CB should become more important with time as the ratio of dissolved anthropogenic carbon per unit mass of DO decreases through engineered remediation and weathering, resulting in increasing DO concentrations across the site. The results obtained from Kelly AFB suggest that many sites with chronic chlorinated benzene contamination may be suited for the application of MNA because indigenous microbial populations can degrade these compounds. Recent releases of CB into an aquifer could experience a lengthy lag time before CB degradation could occur. However, once microbial communities have adapted to the presence of CB, the biodegradation of CB may be limited only by the availability of electron acceptors and the prevailing site geochemistry.

REFERENCES

Bosma, T. N. P., J. R. van der Meer, G. Schraa, M. E. Tros, and A. J. B. Zehnder. 1988. "Reductive Dechlorination of all Trichloro- and Dichlorobenzene Isomers:" *FEMS Microbial Eco*l., v. 53: 223-229.

Buscheck, T.E. and C.M. Alcantar. 1995. "Regression Techniques and Analytical Solutions to Demonstrate Intrinsic Bioremediation." In *Proceedings of the 1995 Battelle International Conference on In-Situ and On Site Bioreclamation.* April.

CH2M Hill. 1997. *Final Draft, Volume I, Kelly Air Force Base, IRP Zone 5, Remedial Investigation Report*. January.

de Bont, J. A. M., M. J. W. Vorage, S. Hartmans, and W. J. J. van den Tweel. 1996. "Microbial Degradation of the 1,3-Dichlorobenzene." *Applied and Environmental Microbiology*, v. 52: 677-680.

Fathepure, B. Z., J. M. Tiedje, and S. A. Boyd. 1988. "Reductive Dechlorination of Hexachlorobenzene to Tri- and Dichlorobenzenes in an Anaerobic Sewage Sludge." *Applied and Environmental Microbiology* v. 54: 327-330.

Haigler, B. E., S. F. Nishino, and J. D. Spain, 1988. "Degradation of 1,2-Dichlorobenzene by a Pseudomonas sp." *Applied and Environmental Microbiology* v. 54: 294-301.

Holliger, C., A. J. M. Schraa, and A. J. B. Zehnder. 1992. "Enrichment and Properties of an Anaerobic Mixed Culture Reductively Dechlorinating 1,2,3-Trichlorobenzene to 1,2-Dichlorobenzene." *Applied and Environmental Microbiology* v. 58: 1636-1644.

Moutoux, D. E., L. A. Benson, T. H. Swanson, T. H. Wiedemeier, J. Lenhart, J. T. Wilson, and J. E. Hansen. 1996. "Estimating the Changing Rate of Anaerobic Reductive Dechlorination of Chlorinated Aliphatic Hydrocarbons in the Presence of Petroleum Hydrocarbons." In *Proceedings of the 1996 API/NGWA Conference on Petroleum Hydrocarbons and Organic Chemicals in Ground Water*. Houston, Texas. November.

Nishino, S. F., J. C. Spain, and C. A. Pettigrew. 1994. "Biodegradation of Chlorobenzene by Indigenous Bacteria." *Environmental Toxicology and Chemistry* 13(6): 871-877.

Parsons Engineering Science, Inc. 1999. *Final Treatability Study in Support of Monitored Natural Attenuation at Site S-1, Zone 5, Kelly AFB, Texas*. Prepared for the Air Force Research Laboratory, Materials and Manufacturing Directorate, Air Base and Environmental Division, Tyndall AFB, FL and SA-ALC/EMRO, Kelly AFB, Texas. February. Denver, CO.

Ramanand, K., M. T. Balba, and J. Duffy. 1993. "Reductive Dehalogenation of Chlorinated Benzenes and Toluenes under Methanogenic Conditions." *Applied and Environmental Microbiology*: 3266-3272. October.

Schraa, G., M. L. Boone, M. S. M. Jetten, A. R. W. van Neerven, P. J. Colberg, and A. J. B. Zehnder. 1986. "Degradation of 1,2-Dichlorobenzene by *Alcaligenes sp.* Strain A175." *Applied and Environmental Microbiology* v. 52: 1374-1381.

Spain, J.C. 1996. "Future Vision: Compounds with Potential for Natural Attenuation." In *Proceedings of the Symposium on Natural Attenuation of Chlorinated Organics in Ground Water*, Dallas, TX, September 11-13, 1996, EPA/540/R-96/509.

Spain, J.C. 1997. "Synthetic Chemicals with Potential for Natural Attenuation." Abstract in *Bioremediation Journal I*(1): 1-9.

Spain, J. C. and S. F. Nishino. 1987. "Degradation of 1,4-Dichlorobenzene by a Pseudomonas sp." *Applied and Environmental Microbiology* v. 53: 1010-1019.

Suflita, J. M. and G. T. Townsend. 1995. "The Microbial Ecology and Physiology of Aryl Dehalogentation Reactions and Implications for Bioremediation." *Microbial Transformation and Degradation of Toxic Organic Chemicals*: 243-268. Department of Botany and Microbiology, University of Oklahoma, Norman, Oklahoma.

US Environmental Protection Agency (USEPA). 1998. *Bioplume III Natural Attenuation Decision Support System, Version 1.0.* Office of Research and Development, Washington, D.C., January, EPA/600/R-98-010.

van der Meer, J. R., C. Werlen, S. F. Nishino, and J. C. Spain. 1998. "Evolution of a Pathway for Chlorobenzene Metabolism Leads to Natural Attenuation in a Contaminated Groundwater." *Applied Environmental Microbiology* v. 64: 4185-4193.

NATURAL ATTENUATION AS A REMEDIAL REMEDY FOR A RECALCITRANT HEXAVALENT CHROMIUM PLUME

Richard S. Makdisi, R.G. .(Stellar Environmental Solutions, Berkeley, California)

ABSTRACT: Groundwater within the shallow (<25 feet bgs) sand aquifer at a Berkeley, California site has become contaminated with Cr(VI) chromium due to release of plating wastes during the 1970s. Groundwater monitoring from 1993 to 1999 shows the predominantly narrow elongated Cr(VI) plume extending 1,200 feet downgradient of the source. Lithologic analyses suggest that the plume's lateral migration is limited by hydrogeologic factors. Total Cr, Cr(VI), dissolved iron, dissolved oxygen, and redox potential sampling data suggest that natural attenuation is occurring, with a significant fraction of Cr(VI) speciating to Cr(III). Selected soil samples from the aquifer material were analyzed for total Cr, Cr(VI), chromium oxidation potential/capacity, manganese (Mn) oxide, soil pH, cation exchange capacity, organic matter content, and particle size. Cr(VI) partition coefficients for the aquifer sediment compare favorably with values derived from the literature, and with retardation factors calculated from observed plume migration rates between 1997 and 2000. Dissolved and adsorbed masses of Cr(VI) were calculated from contoured plume maps using groundwater data collected between 1997 and 2000 to estimate the total mass of Cr(VI) in the aquifer. Comparison of the initial mass calculations in 1997 to the 1999 results show that approximately 41 percent of the Cr(VI) chromium present in the aquifer in 1997 was removed (presumably by reduction) by 1999. Thermodynamic calculations based on platinum electrode redox and pH measurements indicate that solid trivalent chromium hydroxide species are the stable phase in groundwater. Reduction of Cr(VI) to the trivalent state results in precipitation from solution of a solid Cr(III) hydroxide, probably in association with iron oxyhydroxides present in the aquifer. Two possible reducing agents in the shallow groundwater are naturally occurring organic carbon and ferrous iron, both of which were reported in sufficient concentrations in capillary fringe soil sample analyses. Measurements of the rate of Cr(VI) reduction indicate that the reaction proceeds slowly at near-neutral pH values, representative of the aquifer outside the immediate source area. The viability for Monitoring Natural Attenuation (MNA) as a remediation remedy for the chromium plume was tested through the simulation of future plume concentrations using the MYGRT groundwater model (EPRI, 1989). Five years of empirical data were used for the model calibration. The Model predicted attenuation through the year 2035, indicating that the Cr(VI) will be at a concentration of less than the 50 μg/L drinking water standard within 10 to 15 years, or by the time the plume migrates 2,000 feet downgradient of the source. This distance corresponds to approximately 1,000 feet of the San Francisco Bay shoreline. Model predictions were validated with post modeling hydropunch samples, which showed that reported concentrations were slightly less than predicted concentrations near the predicted leading edge of the plume.

INTRODUCTION

The Site facility is located in an industrial area along the western margin of the city of Berkeley, Alameda County, California. The site is an active plating and engraving facility which produces rotors for the printing industry. Chrome, nickel, and copper plating are used in the manufacturing processes. Copper, chromate, and nickel solutions—as well as degreasers, acids, and other chemicals including solvents—were historically used in the plating and engraving process. Leakage of chromium into the shallow groundwater aquifer is traced to 1976, based on the facility history. Until about 1976, the operation used a 6-foot deep dipping tank contained in a concrete over-spill sump, referred to as the former chromium sump. The concrete sump contained water for cooling down the dipping tanks. The concrete over-spill vault was deepened to 10 feet below ground surface (bgs) in order to accommodate larger size drums which resulted in the vault being below the groundwater table. Spillage of the water mixed with chromium solution from the vault was the source of Cr(VI) to the groundwater. The deeper sump was closed in 1993.

WRE/ColorTech has voluntarily pursued the investigation and remediation at the site, with continuous communication with the local lead agency, the City of Berkeley Toxic Management Division (TMD); the Regional Water Quality Control Board (RWQCB) is also taking an active interest. The site facility investigation and completed remediation took place at the site between 1990 and 1999. All primary source areas have been remediated by excavation, including a chromium sump, a xylene sump, three chromate wastewater USTs, associated drain line piping, and contaminated soil. The facility discontinued chemical engraving activity and on-site wastewater treatment in 1993, and all current facility processes are environmentally protective.

Following removal of four redundant groundwater monitoring wells in March 1999, a total of ten wells is now available for monitoring [including one on-site well, four wells immediately downgradient of the site, and five wells wells between 200 and 1,800 feet downgradient (west) of the site]. Historical groundwater monitoring before 1998 analyzed only for total Cr only, rather than total Cr and CR(VI) which allows for the calculation of Cr(III). Trivalent chromium is calculated using the analyzed concentrations of total chromium minus hexavalent chromium. The early signature of the plume was created by the source input into the shallow groundwater regime before the source areas were remediated, migrating downgradient approximately 1,800 feet from the ColorTech facility. Lithologic analyses suggest that the plume's lateral and vertical migration is controlled primarily by hydrogeologic factors. The widest part of the plume is estimated at 400 to 500 feet.

In 1997, the regulatory oversight agency and the previous site consultant were leaning toward consideration of an engineered reactive barrier solution—an iron reactive wall—as a remedy. This was a viable solution, but there had been no systematic evaluation of the geochemical environment, plume attenuation potential, or human health and ecological risks associated with the no remediation option. Cr(VI) is the considered health risk, and is a regulatory red flag, due to its more toxic, mobile, and soluble qualities relative to Cr(III). A review of in situ reduction of Cr(VI) to Cr(III) studies could be used to determine whether the existing plume geometry, concentration gradient, geochemical conditions, and distance to sensitive receptors are amenable to the process of natural in-situ reduction of Cr(VI) to Cr(III). To this end, SES completed a site closure assessment (1998), including hydrochemical trend analyses and data gap identification. Subsequent groundwater monitoring (September 1998 and February 1999) provided the necessary hydrologic and hydrochemical data used to complete an attenuation model (MYGRT, 1989) and RBCA risk assessment for the hexavalent chromium plume. The model and subsequent data collected suggest that the hexavlent plume is attenuating as predicted. Evidence of lithologic controls impact the lateral extent of the plume. Only the upper water-bearing zone—with an average saturated thickness of 12 feet—is affected by the chromium contamintion.

In-situ precipitation of Cr using reduction has been demonstrated to be effective, both by bench scale testing and actual implementation. Rai, et al. (1988) explored in detail the geochemical conditions necessary for efficient Cr reduction; they have explored the efficiency of different Cr reduction reagents suitable for subsurface injection. Rouse and Pyrih (1991) have outlined necessary, preliminary bench-scale testing needed for site evaluation, and the suitability for in-situ Cr reduction. Crosbie, et al. (1993) have described the successful application of in-situ reduction for a site in Pennsylvania. This approach has also been approved for use at a similar site in Turlock, California, which is being overseen by the EPA's Demonstration Program within the Superfund Innovative Technology Evaluation (SITE) Program.

OBJECTIVES

The key objectives of this paper are to answer the following questions:

- Is in-situ Cr reduction and precipitation a potentially applicable remedial alternative for the site?
- Will the in-situ reduction and precipitation of Cr transform the Cr into a phase that is acceptable in terms of risk to human health and the environment, and in achieving the remediation goals required by regulatory agencies given the size of the plume and the distance to the sensitive receptors?
- Do geochemical conditions for the natural sttenuatonas of the Cr(VI) chromium plume exist?
- Is the estimate's concentration of Cr(VI) at the groundwater/Bay interface, located at a distance of 3,000 feet from the Cr(VI) source, lower than the MCL?
- Does the MYGRT model accurately predict the concentration changes in the plume as it migrates toward the Bay?

METHODS

To evaluate the viability of the natural attenuation of hexavalent chromium, SES (1) reviewed published and unpublished work on abatement hexavalent chromium; (2) determined the optimum model to use given the data set and site conditions; and (3) performed a health and ecological risk assessment to determine if the existing and predicted concentrations would be acceptable. MYGRT was selected. The MYGRT model code is a quasi-analytical model based on an advection-dispersion-retardation-decay equation. The processes considered for an inorganic contaminant are advection, dispersion, and linear sorption (retardation). However, for an organic contaminant, advection, dispersion, retardation, and decay are the various processes included in the differential equation. The governing partial differential equation describes a two-dimensional solute transport of a chemical (in this case, chromium) which is appropriate given that only the shallow aquifer is contaminated and no vertical concentration variations exist.

The major assumptions of the model used for this model are: (1) The groundwater seepage velocity remains constant over the distance of a given simulation (Ficks Law is valid to describe a dispersion mechanism); (2) the dispersion coefficients remain constant for a given simulation; (3) sorption is treated as linear equilibrium, partitioning between aqueous and solid phases; (4) transport of a single solute species for a given simulation is adequate (interactions between species are not considered); and (5) first order kinetics adequately simulate solute transformation or decay in the groundwater environment. The decay rate is assumed to be the same for solutes present in either the solid or liquid phases.

The use of a constant seepage velocity, usually set equal to the average value, is a requirement for most analytical models. Except in areas with large vertical gradients or in homogeneous, layered systems with wide variations in hydraulic conductivity, this assumption is a reasonable approximation for screening level calculations. For most solutes in groundwater, linear partitioning is an adequate approximation for simulating equilibrium sorption. At high concentrations, such as may occur in interstitial solution concentrations, a linear representation may not be applicable. The retardation factor used for "metals" should be based on a partition coefficient (K_d) value which is appropriate for the pH, redox conditions, and ionic strength found in the aquifer (Rai and Zachara, 1984). The retardation factor is assumed to be constant for a given simulation.

Because of the formulation of the code's equations, it is not appropriate for simulating storm events or for tracking a wetting front through an unsaturated formation. The code can, however, provide an estimate of transport through the unsaturated zone if one is willing to use the approximation of a constant moisture content (representing a particular set of hydrologic and meteorologic conditions beneath a disposal site). Given that wastes may leach for long periods of time (e.g., tens or even hundreds of years), and that users may be interested in the concentrations at the water table rather than the concentration profile over depth in the unsaturated zone, such approximations can be useful for screening purposes. One could run the code using a range of typical moisture content values to show the variation in concentration versus time due to this factor. In general, the assumptions and limitations of the MYGRT model are likely to over-estimate, rather than under-estimate the concentration of a chemical in groundwater. This is consistent with its typical use as a screening model.

An assumption of the model was that chemical flow was two-dimensional, i.e., in the X- and Y- directions. This assumption was conservative because a greater thickness of the aquifer would permit dispersion in the Z- direction (downward direction), thereby resulting in greater dilution of the chemical before reaching the receptor. The geological, hydrogeological, and chemical parameters used in the model are shown in Table 1.

In addition, a uniform groundwater velocity of 30 m/yr was assumed. This velocity assumption was conservative compared to the reported aquifer test data value of 12.2 m/yr (40 ft/yr). The soil properties were assumed to be homogeneous within the groundwater flow domain of interest. The Cr(VI) source concentration was conservatively chosen to be 1.000 ppm, and the source existed for a period of time ranging from 1978 to 1995. The half-life for Cr(VI) was considered to be 2.5 years (Henderson, 1994).

TABLE 1: Parameters Used as Model Input Data

SOIL PROPERTIES	VALUE	UNITS	REFERENCE
Effective Porosity (υ)	0.3		Site-specific
Dry Bulk Density (P)	1.5	g/cm^3	USEPA (1996)
Groundwater Velocity (Vx)	30	m/yr	Assumed
K_d	0.2	L/kg	Assumed
Longitudinal Dispersion Coefficient (Dx)	30 x D_x	m^2/yr	D_x determined using Xu and Eckstein (1995)
Transverse Dispersion Coefficient	0.1 x 30 x D_x	m^2/yr	Gelhar (1992)
Width of Source (W)	5	Meters	Site-specific
t_{on}	1978		Site-specific
t_{off}	1995.8		Site-specific
Half-life of Cr(VI)	2.5	Year	Henderson (1994)
Source Concentration (C)	1,000	Ppm	Site-specific

The ability of the MYGRT model to accurately predict the concentration of the soluble fraction of chromium at the Bay interface is based mainly on the model's most sensitive site-specific parameters: (1) concentrations of chromium at the monitoring points (empirical data); (2) measured distance to the Bay (empirical data); (3) groundwater velocity (estimated by aquifer step draw-down test with conservative factor added in by SES); and (4) Cr(VI) half life of 2.5 years (validated by field collected data using Henderson data).

RESULTS AND DISCUSSION

Mechanisms of Chromium Reduction and Controlling Processes. The main geochemical mechanisms for reduction of the Cr(VI) plume are oxidation/reduction reactions in soils, precipitation/dissolution reactions in soils, and adsorption/desorption reactions in soils (Palmer and Wittbrodt, 1991). Cr(VI) is a strong oxidant, and is reduced in the presence of an electron donor such as Fe(II). The Cr (VI) state is generally considered to pose the greatest human health risk because it is more toxic, more mobile, and more soluble than Cr(III). Under the commonly encountered pH range of most soils, sediments, and natural waters, the dominant Cr species tend to be:

- Cr(III)—$CrOH_2^+$, $Cr(OH)_3$, and $Cr(OH)_4$
- Cr(VI)—CrO_4 and $HCrO_4$.

The reduction of Cr(VI) by ferrous iron can be described by the overall reaction:

$$Cr(VI) + 3Fe(II) \sim Cr(III) + 3Fe(III)$$

This reaction appears to be appropriate for pH levels below 10 and PO_4 concentrations less than 0.1 mm. After reduction and oxidation, the Cr(III) and Fe(III) formed can hydrolyze, combine, and precipitate in a solid solution series with Fe(III). This reaction was reported to be very rapid (Palmer and Wittbrodt, 1991), with equilibrium being established in less than 5 mm, even in the presence of dissolved oxygen. The solubility of this solid solution is much less (<10-8) than the Cr(III) precipitate alone. The use of Fe(II) sulfate solutions could enhance the reduction of Cr(VI), but the addition of sulfate is probably not significant since it is a dissolved anion in most California aquifers, as well as in the groundwater at the site.

Potential pitfalls with the Cr(VI) reduction to Cr(III) include the possible re-oxidation of Cr(III) to Cr(VI) While many potential mechanisms exist for the reduction of Cr(VI) to Cr(III), the potential mechanisms for the oxidation of Cr(Ill) to Cr(VI) appear to be limited to oxidation by oxygen or by manganese oxides. However, Bartlett and James (1979) found that oxygen does not react appreciably with Cr(III). The presence of Cr(VI) solids could also impact the natural attenuation of Cr(VI), but Rai, et al. (1988) and Palmer and Wittbrodt (1991) concluded that, in most geologic settings with large observed concentrations of Cr (VI), only $BaCrO_4$ or solid solutions with this solid phase were likely to form in soils contaminated with plating type waste.

Contaminant Mass Calculation. The Cr(VI)/Cr(III) mass in the aquifer was calculated by using isoconcentration areas for, 5,000, 1,000, and 500 μg/L. The saturated thickness was taken as a conservative 12 feet. In March 1997, the year that Cr(VI) was available from the more extensive monitoring well array, a total of 266 pounds were calculated. This compares to the 110 pounds of Cr(VI) mass calculated based on the October 1999 data. A decrease of approximately 2.4 times the mass of Cr(VI) in 2.5 years. Cr(III) was calculated at 141 pounds in March 1998, and increased by 2.1 times to 303 pounds in October 1999.

Discussion of Model Results. The uncertainty involved in the initial concentration of the chemical is fairly high. In this analysis, it was assumed that a constant source existed from 1978 to 1995, and that the source concentration was used as a calibration parameter to conservatively predict the measure concentrations in 1998-2000. Hence, this uncertainty is likely to over-estimate the downgradient receptor concentrations because of the conservatism in the model parameter assumptions.

Based on the input parameters used, it appears that the Cr(VI) is likely to attenuate in a reducing environment (DO = 1.2 mg/L, Fe^{2+} = 1.0 ppm, pH = 7-8) concentrations to well below the Cr(VI) Maximum Contaminant Level (MCL) of 50 μg/L prior to reaching the Bay. Table 2 summarizes the comparison between model prediction and measurement. It is evident that the model over-predicts the Cr(VI) concentration at every monitoring well location. The calibrated model is then used to predict the groundwater concentration at the groundwater/Bay interface located 3,000 feet from the source. The concentration at the groundwater/Bay interface during different time periods is shown in Table 3, where it can be seen that the maximum predicted Cr(VI) concentration at the groundwater/Bay interface based on the conservative assumptions made in the model is 2.63E-5 mg/L in the year 2031. This concentration is well below the MCL for Cr(VI).

The attenuation modeling based on the May 1998 sampling event data predicts that the concentration of hexavaent chromium will be just below the 50 μg/L MCL at 1,500 feet from the source, with a decrease to 2.5 μg/L at a distance of 2,000 feet away from the source.

TABLE 2: Comparison of Predicted and Measured Cr(VI) Groundwater Concentrations at Downgradient Wells and Bay in 1998

Location	Measured Concentration (mg/L)	Predicted Concentration (mg/L)
MW-7	3.2	4.4
MW-8	4.4	6.9
MW-12	2.9	4.4
MW-11	1.7	1.81
Bay	-	2.62E-11

Note: MCL for Cr(VI) is 0.05 mg/L and the Continous Concentration for Fresh Water Aquatic Life Protection is 0.01 mg/L for Cr(VI).

TABLE 3: Predicted Future Groundwater Concentrations at Groundwater/Bay Interface

Year	Concentration (mg/L)
2000	2.2E-12
2005	3.66E-9
2010	2.59E-7
2020	1.09E-5
2025	2.07E-5
2030	2.61E-5
2031	2.63E-5
2035	2.37E-5

This model to evaluate the migration of Cr(VI) in the groundwater used a number of theoretical and site-specific conservative assumptions. The measured groundwater velocity was 12.2 m/yr (40 ft/yr), and that used in the model was approximately 30 m/yr (100 ft/yr). The dispersion parameters used in the model was determined using Xu and Eckstein method which would result in a more concentrated plume, unlike the dispersion parameters determined using Gelhar et al., (1992). The transverse dispersivity assumed in the modeling was one-tenth of the longitudinal dispersion. However, the transverse dispersivity recommended by USEPA is 0.33 times the longitudinal dispersivity. This would further dilute the plume due to lateral spreading. Moreover, the model also assumed a source concentration of 1,000 mg/Kg that existed for 18 years (1978 to 1995). The attenuation factor used in this study was obtained from Henderson (1994). A comparison of some of the site parameters for the Trinity Sand Aquifer site (Henderson, 1994) and the present site is shown in Table 2. It is evident that the subsurface condition for this site is under a greater reducing condition when compared to the Trinity Sand Aquifer site. Hence, to be on the conservative side, a value of 2.5 years for Cr(VI) half-life equal to that used in the Trinity Aquifer Site, was used in this modeling analysis.

Mass reduction in Cr(VI) over a 2.5-year period was calculated at a factor of 2.4, from 260 to 110 pounds of Cr(VI) in the aquifer. This agrees well with the 2.5-year half-life in the Trinity Aquifer study. An associated increase in the apparent mass of Cr(III) also occurred from 141 to 303 pounds, supporting the hypothesis that Cr(VI) is precipitating out as Cr(III). The measured parameters supportive of reduction of Cr(VI) to Cr(III) validate the use of the 2.5-year half-life assumption used in the model.

SUMMARY

The MYGRT attenuation model showed that natural attenuation is occurring, with a significant fraction of the original Cr(VI) valence state reducing to the relatively insoluble and less toxic Cr(III) valence state. In addition, the natural attenuation indicators measured at the site were similar to those measured in the groundwater regime at the Trinity aquifer in Texas; that study indicated a half-life of 2.5 years for Cr(VI). The attenuation model predicted that hexavalent chromium will be at a concentration of less than the 50 μg/L drinking water standard by the time the plume migrates to 2,000 feet downgradient of the source, which is 1,000 feet or more before it interfaces with the Bay.

REFERENCES

Bartlett, R.J and B.R. James, 1979. "Oxidation of Cr in Soils." *J. Environ. Qual. 8*:31.

Crosbie, J., R.A. Bown, and S. O'Neil, 1993. *In-situ Treatment of Cr(VI) with an Iron Reduction Process*. Groundwater Technology, Inc.

Eary, L.E. and D. Rai, D, 1988. "Chromate Removal from Aqueous Wastes by Reduction with Ferrous Iron." *Environ. Sci. Tech. 22*:972.

Gelhar, L.W., C. Welty,and K.R. Rehfeldt, 1992. "A Critical Review of Data on Field Scale Dispersion in Aquifer,." *Water Resources Research 28*(7):1955-1974.

Henderson, T., 1994. "Geochemical Reduction of Hexavalent Chromium in Trinity Sand Aquifer," *Ground Water 32*:477-486.

EPRI, 1989. *MYGRT Users Manual*. Electric Power Research Institute. Electric Power Software Center. Dallas, Texas.

Palmer, C.D. and P.R. Wittbrodt, 1991. Processes Affecting the Remediation of Chromium-contaminated Sites. *Environ. Health Perspective 92*:25-40.

Rai, D. and J.M. Zachara, 1988. *Chromium Reactions in Geologic Materials. EA-5741*. Project Number 2485-3. Battelle, Pacific Northwest Laboratories.

Rouse, J. V. and R.Z. Pyrih, 1991. In Place Cleanup of Chromium Contamination of Soil and Groundwater. GEOCHEM, Inc. Austin. TX.

USEPA. 1996. U.S. Environmental Protection Agency. Soil Screening Guidance: Technical Background Document. Office of Emergency and Remedial Response. USEPA, Washington , D.C

Zachara, J.M., D.C. Girvin, R.L. Schmidt, and C.T. Resch, 1987. Chromate Adsorption on Amorphous Iron Oxyhydroxide in the Presence of Major Groundwater Ions. Environ. Sci. Tech. 21:589.

NATURAL ATTENUATION ASSESSMENT OF MULTIPLE VOCS IN A DEEP VADOSE ZONE

David M. Peterson, Michael A. Singletary and James E. Studer (Duke Engineering and Services, Albuquerque, New Mexico)
David R. Miller (Sandia National Laboratories, Albuquerque, New Mexico)

ABSTRACT:

The fate of six volatile organic compounds (VOC) in a 150-meter deep vadose zone was examined in support of a RCRA Corrective Measures Study of the Chemical Waste Landfill at Sandia National Laboratories, Albuquerque, New Mexico. The study focused on the modeling of potential future transport of the VOCs to exposure media upon the completion of two separate voluntary corrective measures - soil vapor extraction and landfill excavation - designed to significantly reduce contaminant levels in subsurface soils. Modeling was performed with R-UNSAT, a finite-difference simulator that was developed by the U.S. Geological Survey. R-UNSAT facilitated a relatively unique and comprehensive assessment of vapor transport because it (1) simulated the simultaneous movement of all six VOCs, taking into account each constituent's diffusion coefficient as affected by its mole fraction within a mixture of chemicals, and (2) permitted simultaneous assessment of risk to human health via volatilization (air) and drinking water (groundwater) pathways. Modeling results suggested that monitored natural attenuation would represent a viable remedial alternative at the landfill after both voluntary corrective measures were completed.

INTRODUCTION

The Chemical Waste Landfill (CWL), shown in Figure 1, is a facility operated by Sandia National Laboratories (SNL) in Albuquerque, New Mexico. From 1962 to 1985, the landfill received a variety of wastes among which were volatile organic compounds (VOC). In 1990, one of the VOCs, trichloroethene (TCE), was discovered at part-per-billion levels in local groundwater, which is currently approximately 150 meters (m) below ground surface. Subsequent investigations of the site, all performed under the auspices of the Resource Conservation and Recovery Act (RCRA), indicated that moisture contents in the vadose zone below the landfill are relatively low, and that downward migration of aqueous-phase contaminants has been and continues to be extremely slow. Alternatively, unsaturated zone vapor transport from liquid sources just a few tens of feet below the base of the landfill to the saturated zone hundreds of feet below was identified as the most likely cause of TCE presence in groundwater.

In recent years, the CWL has undergone two Voluntary Corrective Measures (VCM). The first comprises a soil vapor extraction (SVE) action, which was begun in 1997 and terminated in 1999. The second consists of landfill excavation (LE), a process that began in 1999 and will be completed in 2002.

Though both VCMs appear to be very effective in reducing contaminant levels to non-threatening levels, it is possible that some VOC contamination from past CWL practices will ultimately remain at low levels in the subsurface. Such a possibility was the subject of a preliminary risk evaluation conducted under the early stages of a RCRA corrective measures study (CMS). The purpose of the risk evaluation was to assess the potential for remaining VOCs to influence human exposures through two separate pathways: (1) air inhalation via constituent volatilization, and (2) water ingestion via groundwater pumping. Because post-SVE assessments have shown that multiple volatile organic constituents may continue to exist deep within the subsurface after the LE is finished, the challenge for the preliminary risk study was to provide estimates of the simultaneous transfer of all VOCs to both exposure media. Such estimates were provided using a model of multi-species, vapor-phase transport in the vadose zone. An indirect product of the evaluation was a measure of each VOC's capacity to attenuate naturally.

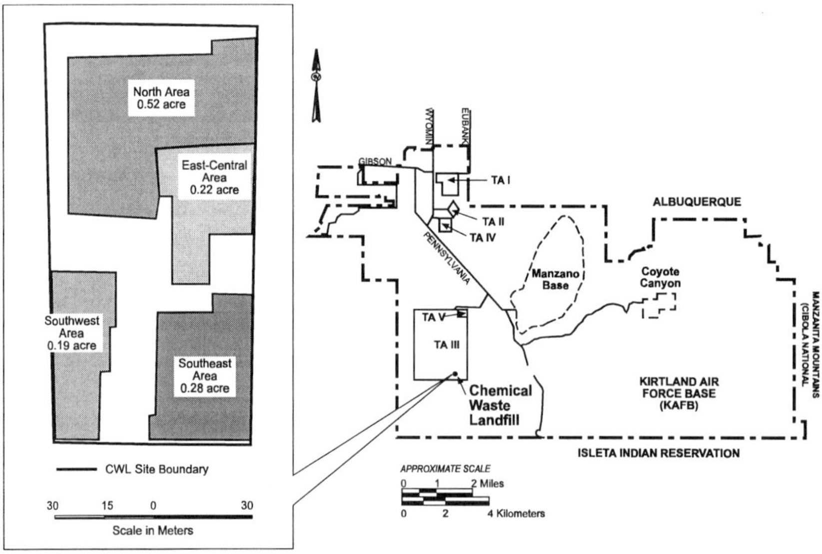

FIGURE 1. Chemical Waste Landfill.

Natural migration of the VOCs was simulated using the finite-difference code R-UNSAT, recently developed by the U.S. Geological Survey (USGS) (Lahvis and Baehr, 1997). Application of R-UNSAT at the CWL accounted for multi-dimensional vapor diffusion and equilibrium partitioning constituent mass among three porous media phases. Several features of the model, such as an accounting of a finite-mass source of contaminants and chemical mixture effects on diffusion coefficients, were especially advantageous for evaluating natural attenuation.

Objective. This study was conducted to provide a preliminary assessment of future health risks stemming from contamination remaining in subsurface soils upon completion of recent and ongoing corrective actions. To make the study as realistic as possible, it was based on risk analysis approaches and techniques discussed in the Soil Screening Guidance (SSG) published by the U.S. Environmental Protection Agency (EPA, 1996a; 1996b). In particular, the modeling performed emulated to the best possible extent the following site-specific conditions: (1) human exposure to contamination at the levels currently observed in remaining subsurface pockets is unlikely (i.e., concentrations are attenuated by transport through unsaturated porous media); (2) remaining subsurface contamination comprises a finite mass; and (3) the apportionment of constituents between competing pathways (i.e., contaminants leaving the subsurface via volatilization are not available to the groundwater pathway).

CONTAMINANT TRANSPORT PROCESSES AT THE CWL

Subsurface transport at the CWL is quite unique because of the 150-m thick unsaturated zone underlying the site. Alluvial sediments in this zone strongly attenuate concentrations of contaminants as they gradually migrate to points of exposure. Of additional significance are the especially dry soil conditions observed in the unsaturated zone, where volumetric moisture contents of about 5 to 12 percent are routinely observed. Such dry conditions are reflective of the very small background infiltration rates of 0.06 to 0.61 cm/yr that have been measured during the SNL Sitewide Hydrogeologic Characterization Project (SNL, 1996). These low rates in turn indicate that, at a minimum, it will take more than 1,000 years for dissolved constituents at landfill level to migrate to the saturated zone. Such hydraulic considerations do not take into account the additional retardation of dissolved chemical movement attributed to sorption on unsaturated sediments. The apparent slow movement of aqueous-phase constituents indicates that the most significant pathways for VOCs to exit the unsaturated zone consist of volatilization across the ground/atmosphere interface and downward vapor transport to the water table. At the latter interface, contamination transfers to groundwater via phase partitioning governed by each contaminant's Henry's Law constant. Once in groundwater, the constituent concentrations are diluted due to mixing of contaminated water with clean water.

Recent post-SVE monitoring identified a vapor plume centered at a depth of about 80 m below ground surface near the CWL's southeast corner. The plume contained six VOCs: TCE, toluene, Freon 113, methylene chloride, trichloroethane (TCA), and 1,1 dichloroethene (1,1 DCE). Observed concentrations of the constituents indicated that no immiscible-phase liquids were present. Thus it could be assumed that there are no continuing VOC sources in the plume other than those stemming from associated aqueous-phase and sorbed contaminants. Accordingly, future unsaturated zone transport of these chemicals will lead to attenuation of contaminant mass and concentrations by losses to the atmosphere and the saturated zone.

R-UNSAT MODEL

The multi-species transport code R-UNSAT (Lahvis and Baehr, 1997) is based on algorithms that incorporate Fick's laws of multi-dimensional vapor flux and assume equilibrium partitioning of constituent mass among the vapor, aqueous, solid, and immiscible phases. Simultaneous, one-dimensional (1-D) downward transport of constituents in the aqueous phase is also simulated with the code. To account for transport of organic constituents below the southeast corner of the CWL, R-UNSAT was used to simulate two-dimensional, axi-symmetric (r,z) vapor diffusion in a domain affected by sediment layering and depth-dependent sediment properties. A domain encompassing 200 m in the radial direction and 150 m in the vertical direction was used for the CWL model. The distance between adjacent nodes in the horizontal (r) direction was set at a uniform value of 2 m, and in the vertical direction, at 3 m. Input to the R-UNSAT model consisted of both physical and chemical descriptors

Alternative means of representing contaminant sources were available in R-UNSAT to selectively characterize any existing contaminant sources. However, because the observed subsurface concentrations of organic constituents at the site indicate that none of them exists in the immiscible phase, R-UNSAT was operated in a manner that omitted continuing contaminant sources. This meant that the model would simulate the eventual, natural fate of the contaminants, whose initial distributions were defined by recently observed subsurface concentrations. Boundary conditions consisted of (1) a prescribed zero concentration along the top of the model (ground surface) where volatilization losses occur and (2) impermeable boundaries along the model domain base (water table) and the outermost radial boundary.

Model Physical Properties. Six separate sediment layers were assigned to the model domain used to represent the plume beneath the CWL's southeast corner. The layer thicknesses and their physical properties were selected to reflect conditions commonly observed in lithologic logs recorded for wells drilled in the east portion of the landfill. In general, these conditions reflect the tendency of sediments to become finer-grained and less permeable with depth. The most notable transitions from shallower materials, which are dominated by relatively coarse-grained sands, to clays and clayey sands, are typically observed at 70 to 82 m below ground surface. Table 1 lists the layer-specific physical properties that were adopted in R-UNSAT; the bottom of the first layer represents the top of the saturated zone, and the top of the sixth layer represents ground surface. An additional domain property affecting transport of VOCs was the downward seepage rate of soil moisture, which was assigned a uniform rate of 0.33 cm/yr.

Chemical-Specific Parameters. Chemical-specific parameters required by R-UNSAT included molecular weight and atomic diffusion volume, which were used to calculate each chemical's binary diffusivity (Lahvis and Baehr, 1997)

$$d_{ki} = \frac{0.00143\,T^{1.75}}{P\,(M_g)^{0.5}\,(\zeta_i^{1/3} + \zeta_k^{1/3})} \tag{1}$$

where d_{ki} = vapor-phase diffusivity of the kth constituent in a binary mixture of the ith and kth constituents (cm^2/sec),
T = temperature (°K),
P = pressure (atmospheres),
ζ_i, ζ_k = atomic diffusion volumes of the ith and kth constituents (unitless),
M_g = $2(\omega_i^{-1} + \omega_k^{-1})^{-1}$, and
ω_i, ω_k = molecular weights of the ith and kth constituents (g).

TABLE 1. Physical Properties of R-UNSAT Model Layers

Layer	Top Elevation (m)	Thickness (m)	Porosity (unitless)	Moisture Content (unitless)	Bulk Density (g/cm^3)
1	3	3	0.4	0.2	1.65
2	27	24	0.4	0.15	1.65
3	51	24	0.4	0.12	1.65
4	60	9	0.35	0.07	1.65
5	75	15	0.4	0.2	1.65
6	96	75	0.35	0.07	1.65

The diffusion coefficient for a constituent in a mixture of N vapor-phase components was determined using the formula (Lahvis and Baehr, 1997)

$$D_{a,i} = \frac{1 - \chi_i}{\sum_{i=1, i \neq k}^{N} \chi_i / d_{ki}} \tag{2}$$

where χ_i = mole fraction of the ith constituent in the vapor phase.
Accounting for each chemical's relative concentration per equation 2 helped to avoid overprediction of volatilization fluxes (Benson et al., 1993) that can occur with simpler models of a single constituent.

To derive appropriate diffusion coefficients per equations 1 and 2, an ambient temperature of 20°C (293°K) and a total vapor-phase pressure of 0.82 atmospheres were assumed. Within the model, the bulk-vapor diffusion coefficient for each VOC was also modified by a tortuosity parameter reflective of the convoluted path that each passes through in porous media. This parameter was calculated in the R-UNSAT code using the porosity and moisture content data from Table 1 in a formula developed by Millington (1959).

Additional chemical-specific parameters used in the model included a Henry's Law constant, H (unitless), and a soil-water partition coefficient, K_d (L/kg), for each chemical. The former expresses the equilibrium ratio of a constituent's vapor concentration to its aqueous-phase concentrations, whereas the latter comprises the ratio of a chemical's concentration in the sorbed phase (mg/kg) to its dissolved concentration in the aqueous phase (mg/L). K_ds were computed as the product of a soil-water partition coefficient, K_{oc} (L/kg), and a soil organic carbon content, f_{oc}, the latter of which was assigned a uniform value of 0.001 (0.1%). Values of H and K_{oc} were estimated using data provided in the Technical Background Document (TBD) of EPA's SSG (EPA, 1996b). Aqueous-phase diffusion coefficients, D_w, were also considered chemical-specific parameters; however, each constituent was assigned a constant D_w of 1×10^{-5} cm^2/sec since this parameter is typically two orders of magnitude less than a comparable vapor diffusion coefficient and thus not expected to strongly affect simulation results.

The values assigned to all model-related chemical-specific descriptors for the six organic chemicals are listed in Table 2. An ancillary parameter, chemical solubility in water, is also listed to provide some measure of each constituent's proclivity for partitioning into soil moisture versus the soil and vapor phases. Because each VOC had non-zero Henry's Law constants and soil-water partition coefficients, its vapor transport was retarded by partitioning of contaminant mass into the aqueous (soil moisture) and solid (sediment) phases. An effective retardation factor for each chemical was calculated in R-UNSAT.

TABLE 2. Chemical-Specific Parameters for VOC Plume at the CWL

Constituent	Molecular Weight ω (g)	Atomic Diffusion Volume ζ (unitless)	Henry's Law Constant, H (unitless)	Organic Carbon /Water Coefficient K_{oc} (cm^3/g)	Solubility (mg/L)
1,1 Dichloroethene	96.94	78.42	1.07	31.6	2250
Freon 113	187.4	138.9	13.08	389	170
Methylene Chloride	84.93	62.52	0.0898	11.7	13000
Toluene	92.14	111.5	0.272	182	526
1,1,1 Trichloroethane	133.4	101.7	0.705	110	1330
Trichloroethene	131.4	97.11	0.422	166	1100

RESULTS AND CONCLUSIONS

Using recently observed vapor concentrations centered at a depth of approximately 80 m as initial conditions, the model was used to simulate a period of 100 years. A sequence of cross-sectional snapshots of computed TCE concentrations from three years through 30 years is presented in Figures 2(a) through 2(d). Temporal plots for the next 40 years of (1) TCE concentration just below ground surface and at the water table; and (2) cumulative TCE mass within the unsaturated zone and exiting the model domain at ground surface (volatilization) are shown in Figures 3(a) and 3(b), respectively. The snapshots clearly demonstrate the capacity of the large unsaturated zone to decrease concentrations with time. As shown by the temporal plots, TCE is expected to begin releasing to the atmosphere via volatilization approximately three years beyond zero time. However, concentrations of this constituent near ground surface are relatively small until 10 years, and begin decreasing after this time. In contrast to TCE levels near the ground surface, notable concentrations of this constituent do not appear at the water table until about eight years of simulation time, and are still increasing after 30 years. Simulation results for the remaining VOCs were similar to those for TCE. However, two of the constituents, Freon 113 and 1,1 DCE, with relatively high Henry's Law constants of 13.08 and 1.07, respectively, were observed to volatilize and reach the saturated zone much earlier than TCE. Similarly, computed vapor concentrations of these more volatile constituents at the water table began decreasing after 14 years of simulation time.

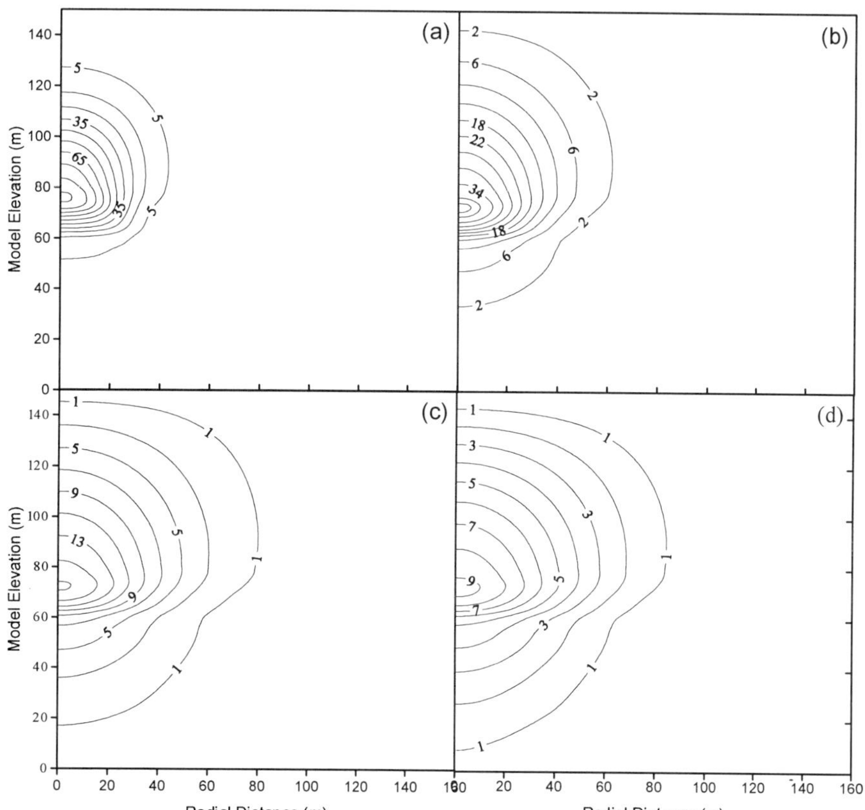

FIGURE 2. TCE concentrations (ppm$_v$) at (a) 3, (b) 10, (c) 20, and (d) 30 yrs.

Preliminary Risk Evaluation. The volatilization pathway was evaluated using model-computed average vapor fluxes of each constituent over the first 30 years of simulation. These fluxes were applied in a relatively simple model (EPA, 1996b) for assessing the impacts of dispersion on emission fluxes to the atmosphere. The resulting calculations indicated that emission fluxes would be relatively insignificant and would thus not create unacceptable risks due to air inhalation. To analyze the effects of vapor diffusion on groundwater, the largest vapor concentrations occurring over time at the top of the saturated zone were translated into equivalent groundwater concentrations. This was accomplished by assuming that each chemical would partition into groundwater in accordance with its Henry's Law constant, and would subsequently be diluted by a default factor of 20 due to mixing with uncontaminated groundwater (EPA, 1996a; 1996b). For each constituent, the resulting computed groundwater concentration was less than the chemical's maximum contaminant level (MCL).

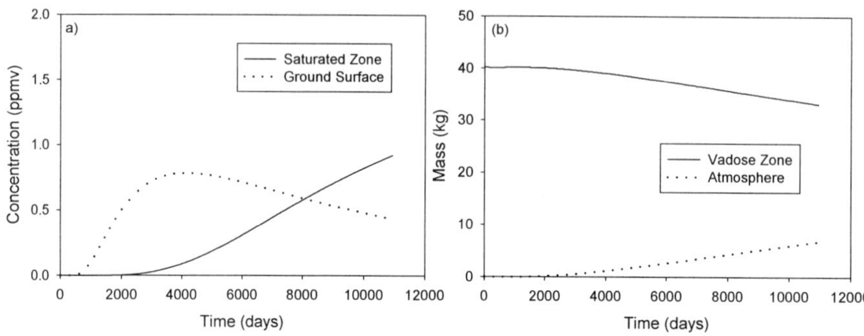

FIGURE 3. (a) TCE concentrations near the ground surface and at the water table. (b) TCE mass in the vadose zone and the atmosphere.

This investigation demonstrated that the simultaneous, eventual fate of multiple VOCs remaining in a deep vadose zone after site remediation could be examined with a numerical model for insight into the contaminants' potential effects on two risk exposure pathways. The 150-m deep vadose zone in this study appeared to be large enough to both minimize risk measures at key exposure points and effectively attenuate VOC concentrations to non-threatening levels.

ACKNOWLEDGMENTS

This study was supported by the U.S. Department of Energy under Contract DE-AC-94AL85000. The views and opinions of the authors do not necessarily state or reflect those of the U.S. Government or any agency therof.

REFERENCES

Benson, D.A., D. Huntley, and P.C. Johnson, 1993. "Modeling Vapor Extraction and General Transport in the Presence of NAPL Mixtures and Nonideal Conditions." *Ground Water*, 31(3): 437-445.

Lahvis, M.A., and A.L. Baehr, 1997. *Documentation of R-UNSAT, A Computer Model for the Simulation of Reactive, Multispecies Transport in the Unsaturated Zone.* U.S. Geological Survey Open-File Report 97-630.

Millington, R.J., 1959. "Gas diffusion in porous media." *Science* (130):100-102.

Sandia National Laboratories (SNL), 1996. *Site-wide hydrogeologic characterization report, calendar year 1995, annual report.*

U.S. Environmental Protection Agency (EPA), 1996a. *Soil screening guidance: user's guide, second edition.* Publication 9355.4-23.

U.S. Environmental Protection Agency (EPA), 1996b. *Soil screening guidance: technical background document.* Publication 9355.4-17A.

NATURAL ATTENUATION MECHANISMS OF INORGANIC POLLUTANTS: SORPTION/DESORPTION KINETICS ON SOIL MINERAL SURFACES

Paul R.Grossl (Utah State University, Logan, Utah)
Warren D. Brady (IT Group, Baton Rouge, Louisiana)
Matthew J. Eick. (Virginia Polytechnic Institute and State University, Blacksburg, Virginia.)

ABSTRACT: Recent advances in soil metal remediation have incorporated corrosive soil leaches, vitrification technologies, and grout curtains to liberate or immobilize inorganic contaminates from soils. Natural attenuation, an alternate *in-situ* approach, may prove to be more cost-effective and provide greater environmental protection in the long run these current technologies. Kinetics of sorption/desorption reactions on soil mineral surfaces, provides information on inorganic ion sorption reaction rates and mechanisms and determines how constituents, such as dissolved organic carbon (DOC), affect reaction rates and mechanisms. This is information that can be used to assess natural attenuation of inorganic pollutants. Adsorption/desorption reactions of inorganic ions on oxide surfaces are characterized by very rapid reaction rates. Relaxation techniques, such as pressure-jump, allow one to measure reaction rates on millisecond time scales. P-jump studies indicated that metal ions such as Cu^{2+}, Pb^{2+}, and Ni^{2+}, and the oxyanions, arsenate and phosphate, formed inner-sphere surface complexes with goethite. The desorption step was the slower process and could be considered the rate-limiting step, which indicates that the inorganic ion-goethite surface bond created by these ions is more difficult to break than to form. Batch studies reveal that DOC can inhibit the precipitation and adsorption of phosphate onto soil mineral surfaces. This process may also affect the sorption of other inorganic ions onto soil surfaces. Inhibition of sorption results in more inorganic pollutants present in the soil solution, which enhances their mobility toward ground water.

INTRODUCTION

The potential for ground and surface water contamination by inorganic pollutants, has generated public and political concern, demanding swift and proper remediation of contaminated sites. Trace element contaminants include lead (Pb), cadmium (Cd), copper (Cu), nickel (Ni), arsenic (As) and phosphorus (P). In natural waters under aerobic conditions, Pb, Cd, Cu, and Ni are present as divalent metal ions, while As and P are present as the arsenate (AsO_4^{3-}) and phosphate (PO_4^{3-}) oxyanions. Sources of contamination are predominately anthropogenic, arising from the disposal of industrial wastes, sewage sludges, mining and smelting operations, and agricultural chemicals. Many current clean-up technologies for inorganic wastes, such as soil extraction and leaching, are costly and ultimately produce another waste product that must be handled. Natural attenuation, an alternate *in-situ* approach, may prove more cost-effective and provide greater environmental protection in the long run these current technologies. Natural attenuation involves the degradation, dispersion, dilution, sorption, volatilization, and/or chemical stabilization of contaminants to reduce contaminant toxicity, mobility, or volume to levels that are protective of human health and the ecosystem.

In natural systems, sorption processes regulate the activity of inorganic ions. Sorption encompasses both adsorption/desorption and precipitation/ dissolution processes. Here, adsorption refers to the net accumulation of contaminant ions in a two dimensional molecular arrangement at a solid/aqueous interface, while precipitation involves the formation of a three dimensional solid phases arrangement of ions from solution. Sorption of inorganic constituents is influenced by their solution concentrations. At lower ion concentrations, undersaturated with respect to relevant solid phases, adsorption processes dominate. Iron oxides solid phases, such as ferrihydrite and goethite, act as major adsorbents in soil and sediments. The adsorption of metals and oxyanions on iron oxide surfaces is pH dependent. Adsorption of metals is increases with increasing pH, while the adsorption of oxyanions is favored under acidic conditions. At higher ion concentrations, metal ions most often precipitate as hydroxide, and carbonate solid phases (Zachara et al., 1989a, b). Depending on the environment to which phosphate waste is added, it tends to precipitate either as calcium or iron phosphates (Lindsay, 1979; Grossl and Inskeep, 1991, 1992). Precipitate formation is also pH dependent.

The time scales for sorption processes in soils vary considerably. For example, surface adsorption/desorption reactions are rapid occurring within milliseconds, whereas, precipitation/dissolution reactions are much slower spanning time scales ranging from minutes to several years (Amacher, 1991). Methods to evaluate the kinetics of sorption processes are designed to measure reaction rates over the time scales for which they occur. Pressure jump (p-jump) relaxation has been successfully used to measure the rapid kinetics of ion surface adsorption reactions on iron oxide surfaces (Grossl et al., 1994, 1997). This method is based on the fact that the equilibrium constant for a chemical reaction is dependent upon pressure. Hence, a pressure perturbation will shift the equilibrium state of a chemical reaction and the system must relax to the final equilibrium state. The relaxation is a function of all the elementary reaction steps that comprise the chemical reaction. Batch techniques are often used to measure the slower kinetics associated with precipitation/dissolution reactions. In general, kinetic studies not only provide information on reaction rates, but also provide mechanistic information.

The sorption of inorganic contaminants on mineral surfaces may be restricted by the presence of other soluble constituents (e.g., organic matter). Soluble organic matter is ubiquitous in most aqueous systems DOC levels ranging from about 1 to 30 mg/L for ground waters, streams, rivers and the interstitial waters of soils (Thurman, 1985). A significant portion of the DOC of these waters is comprised of humic and fulvic acids. Humic and fulvic acids are known to sorb onto and, in many cases, coat the surfaces of soil solids, including Fe oxides (Thurman, 1985; Jardine et al., 1989). Thus, soluble organic matter may compete with inorganic ions for sorption sites on solid surfaces and restrict the retention of pollutants in soil systems. This may place limitations on natural attenuation of inorganic wastes.

Our objective is to use chemical kinetics to provide information on inorganic ion sorption reaction rates and mechanisms and determine how constituents, such as DOC, affect reaction rates and mechanisms. This is information that can be used to assess natural attenuation of inorganic pollutants.

MATERIALS AND METHODS

Pressure-jump relaxation kinetic experiments were conducted using a p-jump apparatus manufactured by DIA-LOG (Duesseldorf, Germany). The apparatus consisted of a conductivity detection unit, a pressure autoclave, and sample and reference electrode cells. Prior to conducting p-jump experiments, suspensions containing iron oxides and the chosen sorbate ion were equilibrated at predetermined pH, ionic strength, and sorbate concentration conditions. Sample suspensions and reference solutions (background electrolyte solution) were placed into the electrode cells and both were inserted into the pressure autoclave. The cells contained small electrodes that monitored conductivity changes occurring during a relaxation experiment. Pressure was applied to the sample suspension and reference solution within the electrode cell via the pressure autoclave. The autoclave chamber was sealed with a thin strip of brass foil that was milled to burst once the pressure within the chamber reached 13.5 Mpa. Thus, the pressure perturbation was achieved by increasing the pressure, using a hand-operated pump that forced water into the chamber, until the brass foil burst and the pressure instantaneously dropped to ambient conditions. The sample equilibrated at high pressure relaxed to the new equilibrium obtained at ambient pressure. The pressure changes were transmitted to the sample and reference cells via a thin Teflon membrane that covered and enclosed the sample suspensions and reference solutions within the cells. The relaxation information was monitored by the conductivity detection unit comprised of a wheatstone bridge; where two arms of the bridge included the sample and reference electrodes and the other two arms included variable resistors and capacitors. The bridge was balanced by the variable resistors and capacitors under ambient conditions and became unbalanced at higher pressure. After the brass foil burst a piezoelectric capacitor triggered the collection of the relaxation event, recorded as the voltage change associated with the bridge returning to the balanced condition set under ambient conditions.
This information was digitized and a software program provided by the manufacturer computed the relaxation time constants (tau,) from the digitized information. Tau was the time it took the relaxation to reach 1/e of the initial relaxation amplitude. A more detailed description of p-jump experiments is found in Grossl et al. (1994).

Batch precipitation and adsorption kinetic studies were conducted in a flat-bottomed, water-jacketed reaction vessel (400-600 mL) covered with a removable Plexiglas lid containing entry ports for a stirrer, a pH electrode, nitrogen gas, and burette tip. Experiments were run with and without DOC, added either as humic or fulvic acids obtained from the International Humic Substances Society. For adsorption experiments, the inorganic ion of interest (i.e., Cu, arsenate, phosphate) was added to a given iron oxide suspension with a constant background ionic strength. A pH-stat titrator maintained a constant pH within the reacting solutions by adding either strong acid or base and equilibrium was reached when no more acid or base was added. Sample aliquots were collected during experiments and ion concentrations analyzed to determine the amount of the chosen ion lost from solution. For precipitation experiments, a seeded crystal growth technique was used (Grossl and Inskeep; 1991, 1992). Here solutions supersaturated with respect to a given solid phase were allowed to stabilize prior to the addition of crystal seeds. The addition of crystal seeds triggered the precipitation reaction. Again, sample aliquots were collected during experiments and ion concentrations analyzed either by atomic absorption (AA) spectrometry or inductively coupled plasma (ICP) spectrometry, to determine the amount of the chosen ion lost from solution.

RESULTS AND DISCUSSION

Results from pressure-jump (p-jump) relaxation kinetic experiments evaluating Cu^{2+} adsorption/desorption on goethite indicate that Cu^{2+} was specifically adsorbed to the goethite surface forming a monodentate inner-sphere surface complex (Grossl et al., 1994). The rate constant for the desorption step ($10^{4.88}$ L $mol^{-1}s^{-1}$) was two orders of magnitude smaller than that for adsorption step ($10^{6.81}$ L $mol^{-1}s^{-1}$). Hence, desorption was the slower process and could be considered the rate-limiting step, indicating that the Cu-goethite surface bond is more difficult to break after it is formed. Using data from Pb^{2+} (Hayes, 1987) and Ni^{2+} (Mehadi, 1993) adsorption/desorption on goethite (both of these metal ions formed inner-sphere complexes with goethite), Grossl et al.(1994) found that the rate of adsorption of a divalent metal ion was related to the rate of release of water from the primary hydration sphere of a given divalent metal. As a result, if the detachment rate of a water molecule is slow then the adsorption of the metal ion will also be slow.

P-jump studies revealed that arsenate was specifically adsorbed on goethite following a two step reaction mechanism (Grossl et al., 1997). The first step involved the formation of an inner-sphere, monodentate arsenate-goethite surface complex, which in the second step proceeded to form an inner-sphere bidentate surface complex. Overall, the rate constants for desorption were nearly five orders of magnitude slower than for adsorption, hence, desorption was the rate limiting step. The results are consistent with X-ray absorption spectroscopic studies conducted by Waychunas et al. (1993) and Fendorf et al. (1997) indicating that arsenate forms predominately an inner-sphere bidentate surface complex with goethite and other iron oxides. Recent work by Eick and Tadanier (unpublished data) indicates that phosphate sorption on iron oxides behaves similarly to arsenate. Although the above studies only investigated surface adsorption reactions, where adsorption implies only monolayer surface coverage, sorption processes may actually involve a continuum between surface complexation and precipitation reactions (Farley et al., 1985).

In natural systems the sorption of arsenate on iron oxides may be restricted by DOC present as humic and fulvic acids. Humic substances can sorb onto and coat the surfaces of soil solids, including Fe oxides (Thurman, 1985; Jardine et al., 1989). Humic substances have been shown to inhibit the sorption of phosphate, an oxyanion chemically similar to arsenate, on mineral surfaces (Grossl and Inskeep, 1991).

Eutrophication of water bodies is accelerated by the amount of bioavailable P present in runoff waters. Bioavailable P consists of soluble phosphate in the runoff waters and bioavailable phosphate associated with runoff particles (Sharpley, 1993). The soluble phosphate is immediately available to algae whereas the particle bound phosphate comprises a long-term source of bioavailable phosphate (Sharpley et al., 1992). Ultimately, phosphate sorption/desorption on soil surfaces is the factor controlling soluble phosphate levels in natural waters. In calcareous systems, with pH values ranging from near neutral and above, dicalcium phosphate dihydrate (DCPD), octacalcium phosphate (OCP), and hydroxyapatite (HAP) are major P solid phases (Lindsay, 1979). Thermodynamically, HAP is expected to control phosphate levels in soil solutions (Lindsay, 1979), however, more soluble phases such as DCPD, OCP, and surface adsorbed phosphate frequently control phosphate levels for considerable lengths of time (Fixen et al., 1983; O'Connor et al., 1986). The metastability of the more soluble solid phases with respect to HAP, whose formation is relatively slow, suggests that kinetics play an important role in regulating phosphate levels in natural waters. Calcium phosphate precipitation can be conceptually explained using principles of chemical kinetics and reaction pathways (Grossl and Inskeep, 1992). Starting with a solution supersaturated with respect to DCPD, OCP, and HAP, precipitation can proceed either through a parallel or series reaction pathway. In either case, DCPD would precipitate first, and since the formation of OCP and HAP are inhibited to a greater extent by DOC, present at levels realistic for many soil environments, DCPD would remain as a metastable solid phase controlling phosphate levels in soil solutions.

Dissolved organic carbon can also compete with phosphate for sorption sites on oxide surfaces (Mackowiak and Grossl, 1999). We found that humic acid inhibited the adsorption of phosphate onto the iron oxide ferrihydrite by 43 % Figure 1). We surmise that the hydrophobic nature of the humic acid lead to an energetically favorable entropy-driven adsorption process, that resulted in coverage of surface sorption site. This surface "poisoning" temporarily hindered the adsorption of phosphate on ferrihydrite (Mackowiak and Grossl, 1999).

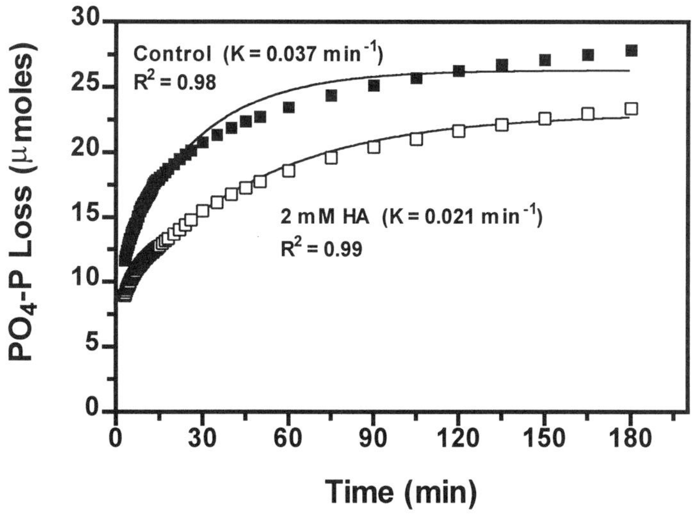

FIGURE 1. Loss of phosphate from solution, due to adsorption onto ferrihydrite, with and without 2 mM humic acid.

CONCLUSIONS

Kinetic studies can be used to determine reaction rates and mechanisms. P-jump studies revealed, that metal ions such as Cu^{2+}, Pb^{2+}, and Ni^{2+}, formed inner-sphere surface complexes with goethite. The desorption step was the slower process and could be considered the rate-limiting step. This indicates that the metal ion-goethite surface bond created by these divalent metal ions is more difficult to break than to form. Arsenate was specifically adsorbed on goethite following a two-step reaction mechanism. The first step involved the formation of an inner-sphere, monodentate arsenate-goethite surface complex, followed by the second step, the formation of an inner-sphere bidentate surface complex. Overall, the rate constants for desorption were nearly five orders of magnitude slower than for adsorption, and, again desorption was the rate limiting step.

Dissolved organic carbon can inhibit the precipitation and adsorption of phosphate onto soil mineral surfaces. DOC poisons surface sorption reactions by blocking active sorption sites. This process may also affect the sorption of other inorganic ions onto soil surfaces, but further study is needed in this area. Ultimately, inhibition of sorption results in more inorganic pollutants present in the soil solution, which enhances their mobility toward ground waters.

The significance of kinetic studies is that they help predict the sorption behavior and fate of polluting inorganic ions in natural systems. This is necessary information for proper natural attenuation of inorganic contaminants.

REFERENCES

Amacher, M.C. 1991. "Methods of Obtaining and Analyzing Kinetic Data." p. 19-59. In: D.L. Sparks and D.L. Suarez (eds.) *Rates of soil chemical processes.* SSSA Special Publ. No. 27. Soil Science Society of America, Madison, WI.

Farley, K.J., D.A. Dzombak, and F.M.M. Morel. 1985. "A Surface Precipitation Model for the Sorption of Cations on Metal Oxides." *Journal of Colloid and Interface Science.* 106:226-242.

Fendorf, S.E., M.J. Eick, P.R. Grossl, and D.L. Sparks. 1997. "Arsenate and Chromate Retention Mechanisms on Goethite: I. Surface Structure." *Environ. Sci. Technol.* 31:315-320.

Fixen, P.E.., A.E. Ludwick, and S.R. Olsen. 1983. "Phosphorus and Potassium Fertilization of Irrigated Alfalfa on Calcareous soils: II. Soil Phosphorus Solubility Relationships." *Soil Sci. Soc. Am. J.* 47:112-117.

Grossl, P.R., and W.P. Inskeep. 1991. "Precipitation of Dicalcium Phosphate Dihydrate in the Presence of Organic Acids." *Soil Sci. Soc. Am. J.* 55:670-675.

Grossl, P.R., and W.P. Inskeep. 1992. "Kinetics of Octacalcium Phosphate Crystal Growth in the Presence of Organic Acids." *Geochimica et Cosmochim. Acta.* 56:1955-1961.

Grossl, P.R., D.L. Sparks, and C. Ainsworth. 1994. "Pressure-jump Kinetic Study of Cu(II) Adsorption/Desorption on Goethite." *Environ. Sci. & Technol.* 28:1422-1429.

Grossl, P.R., M.J. Eick, D.L. Sparks, S. Goldberg, and C. Ainsworth. 1997. "Arsenate and Chromate Retention Mechanisms on Goethite. II. Kinetic Evaluation

Using a Pressure-jump Relaxation Technique." *Environ. Sci. & Technol.* 31:321-326.

Hayes, K.F. 1987. "Equilibrium, Spectroscopic, and Kinetic Studies of Ion Adsorption at the Oxide/Aqueous Interface." Ph.D. dissertation. Stanford University, Palo Alto, California.

Jardine, P.M., N.L. Weber, and J.F. McCarthy. 1989. "Mechanisms of Dissolved Organic Carbon Adsorption on Soil." *Soil Sci. Soc. Am. J.* 53:1378-1385.

Lindsay W.L. 1979. *Chemical equilibria in soils.* J. Wiley & Sons, New York.

Mackowiak, C., and P. Grossl. 1999. "Iron Phosphate Precipitation in the Absence and Presence of Soluble Organic Matter." 1999 Annual Meeting Abstracts. American Society of Agronomy 91st Annual Meeting. p.317.

Mehadi, A. 1993. "Reaction of Ni with Soil and Goethite: Equilibrium and Kinetic Studies." Ph.D. dissertation. University of New Hampshire, Durham, New Hampshire.

O'Connor, G.A., K.L. Knudtsen, and G.A. Connel. 1986. "Phosphorus Solubility in Sludge-amended Calcareous Soils." *J. Environ. Qual.* 15:308-312.

Sharpley, A.N. 1993. "An Innovative Approach to Estimate Bioavailable Phosphorus in Agricultural Runoff Using Iron Oxide-impregnated Paper." *J. Environ. Qual.* 22:597-601.

Sharpley, A.N., S.J. Smith, B.A. Stewart, and A.C. Mathers. 1992. "The Transport of Bioavailable Phosphorus in Agricultural Runoff." *J. Environ. Qual.* 21:30-35.

Thurman, E.M. 1985. *Organic Geochemistry of Natural Waters.* Nihjoff/Junk Publ., Dordrecht, the Netherlands.

Waychunas, G.A., B.A. Rea, C.C. Fuller, and J.A. Davis. 1993. "Surface Chemistry of Ferrihydrite: Part 1. EXAFS Studies of the Geometry of Coprecipitated and Adsorbed Arsenate." *Geochim. Cosmochim. Acta.* 57:2251-2269.

Zachara, J.M., J.A. Kittrick, and J.B. Harsh. 1989a. "The Mechansim of Zinc Adsorption on Calcite." *Geochem. Cosmochim. Acta.* 52:2281-2291.

Zachara, J.M., J.A. Kittrick, L.S. Dake, and J.B. Harsh. 1989b. "Solubility and Surface Spectroscopy of Zinc Precipitates on Calcite." *Geochem. Cosmochim. Acta.* 53:9-19.

EVALUATION AND IMPLEMENTATION OF NATURAL ATTENUATION AT METALS IMPACTED SITES

Warren D. Brady (IT Corporation, Baton Rouge, LA)
Duane Graves and Julia Klens (IT Corporation, Knoxville, TN)
Matthew Eick (Virginia Polytechnical Institute and State University, Blacksburg VA)

Abstract: The evidence supporting monitored natural attenuation (MNA) of metals has been widely demonstrated. Nonetheless, MNA has only recently been applied effectively for metal impacted sites. MNA of metals requires sequestering or valence state changes by the organic and inorganic components of the soil matrix. This presentation provides a methodology for 1) screening metals impacted sites, 2) collecting baseline data to characterize metal attenuation, and 3) demonstrating attenuation mechanisms for metals in soils and sediments. *MNAtoolbox* (Developed by Sandia National Laboratories) is an effective tool for developing conceptual models for natural attenuation of organic and inorganic constituents. Use of *MNAtoolbox*: 1) identifies likely attenuation pathways, 2) clearly identifies sites where MNA is inappropriate, and 3) evaluates data requirements for subsequent reliance on MNA as a sole or partial corrective action.

Two sites impacted by metals (copper, mercury, lead, nickel, and chromium) were screened using *MNAtoolbox* to determine the potential for MNA and develop a preliminary list of data needs for demonstrating MNA. Sampling and analysis was also performed to characterize metal attenuation mechanisms and develop a MNA program for metals remediation.

INTRODUCTION

Natural attenuation is being increasingly relied upon for the passive remediation of contaminated soils and groundwaters (Brady, et al. 1997). The soil and/or groundwater conditions that favor ambient biodegradation of hydrocarbons and chlorinated organic contaminants is well documented (Rice, 1995; Chapelle, 1996) and generally well accepted. Inorganic contaminants (metals) can also be naturally attenuated through various biological, chemical, and physical processes that result in metal immobilization. Metals often reside in the subsurface for long periods of time where they may become unavailable for biologic uptake (and hence pose less risk) due to sequestration at mineral and organic surfaces, precipitation and/or coprecipitation of solid phases, and redox transformations to a less toxic and less mobile forms. The long-term transport of metals is directly reduced by these processes. When demonstrating natural attenuation of metals in soil, a reduction in bioavailability by these processes must be demonstrated. This paper provides the screening procedures and data collection activities for two metals impacted sites where MNA was evaluated as a remedial alternative.

MNA EVALUATION PROCESS DESCRIPTION

MNA of metals impacted soil and sediments requires a phased approach to properly characterize attenuation mechanisms in a cost-effective manner. Figure 1 provides a step-wise metals MNA evaluation process. The major steps of the MNA evaluation include: 1) preliminary screening, 2) evaluation of data needs, 3) demonstrating MNA, and 4) developing a closure plan.

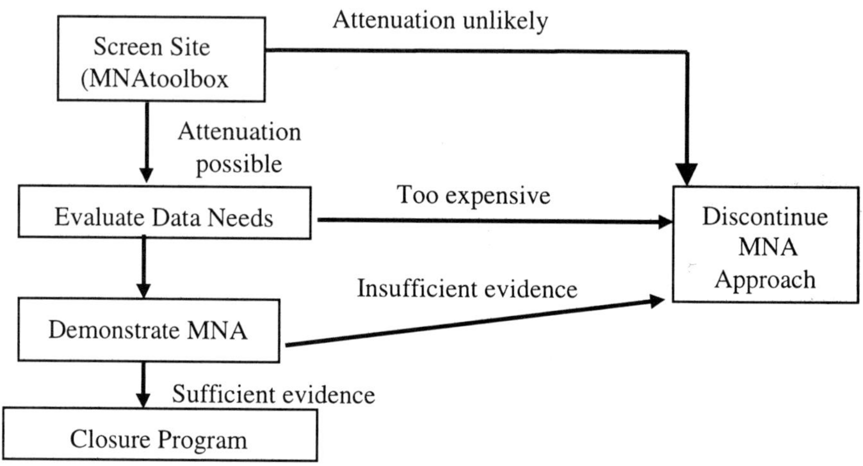

FIGURE 1. MNA evaluation approach for metals.

The first phase of the MNA evaluation is to perform a preliminary screening of the site to determine the likelihood of MNA meeting remedial objectives. Site screening relies on existing characterization data to determine if there is a potential for MNA. This assessment must take into account whether chemical and hydrologic conditions are favorable for MNA to meet remedial objectives prior to impacting a receptor. One rapid means for making such an assessment is to use the *MNAtoolbox*, which is maintained by DOE at www.sandia.gov/eesector/gs/gc/na/mnahome.html (Brady et. al., 1999). MNAtoolbox provides a template for development of the natural attenuation conceptual model.

With an identification of attenuation factors that may contribute to metal immobilization, the MNAtoolbox can be further used to identify the most relevant site characterization data (Step 2) to demonstrate attenuation for site constituents of concern (COCs). From this evaluation, a preliminary assessment of the data needed to demonstrate natural attenuation can be ascertained. The data collection for demonstrating natural attenuation is further refined by addressing regulatory data requirements.

The third step in the process is sampling and analysis to characterize and quantify the applicable attenuation mechanism at the site. If MNA is viewed as a viable option then a closure strategy is provided during the final step. The following section provides a description of the MNA evaluation process and data collection activities used in support of a natural attenuation remedy for two metal impacted sites.

SITE #1- COPPER, CHROMIUM, AND ZINC

Soils from two surface impoundments contained elevated levels of copper, chromium, and zinc. A geochemical evaluation was performed to determine potential metal immobilization processes using *MNAtoolbox*. The *MNAtoolbox* has a scorecard that calculates the potential for implementation of natural attenuation with limited site data. The major attenuation mechanisms identified for these metals were sorption and precipitation/coprecipitation of metals into insoluble metal complexes. Low pH (<4) was noted for some areas of the impoundment. Because pH highly influences metal solubility, a range of pH values were used as inputs in the *MNAtoolbox* scorecard calculation. Scorecard results indicated that MNA was more likely in areas of near neutral pH (>6) than in areas of low pH. Based on an MNA scorecard result that indicated that here was a high potential for implementation in areas of neutral pH area, MNA was investigated. The following subsections provide a description of the methods used to characterize attenuation mechanisms at the site.

Total and Soluble Metal Analysis. Per regulatory requirements, total metals analysis and a deionized water extraction (DIWET) were performed on site samples to estimate the soluble and insoluble fraction of metals. The DIWET method uses a deionized water extraction to determine the fraction of metal that is soluble. Analytical results demonstrated that at near neutral pH the majority of site metals had low water solubility. These results provided indicated that natural processes were immobilizing site metals.

Sequential Extraction Procedures. In addition DIWET extractions, the regulators required an evaluation of the long term stability of metal attenuation mechanisms. A sequential extraction procedure (SEP) was performed to aid in this characterization. SEP procedures generally target the six soil fractions discussed below.

- Exchangeable Phase- metals that are reversibly sorbed to soil minerals, amorphous solids, and/or organic material by electrostatic forces. These forces are overcome by using a concentrated electrolyte leach (1M $MgCl_2$) that displaces the contaminant ions from solid surfaces.
- Carbonate Phase- metals that are irreversibly sorbed or otherwise bound in carbonate minerals can be removed through mild acid dissolution (1M NaOAc solution in 25% HOAc at pH 5).
- Metal Hydroxide Phase - metals bound to hydroxides of iron, manganese, and/or aluminum are extracted using a solution of (0.1 M hydroxylamine hydrochloride) in pH 2 nitric acid.
- Organic Phase - metal associated with soil organic matter (e.g., humic acids) can either be held electrostatically (exchangeable phase) or strongly bound via

chemosorption. Oxidation of soil organic matter (using pH 8.5 5% NaOCl), will release metals that are irreversibly bound to organic ligands.

- Acid/Sulfide Fraction - Any metal associated with sulfide minerals will be removed by leaching with 4M HNO_3 at 95°C.
- Residual Fraction - The remaining metals will be distributed between silicates, phosphates, and refractory oxides and can be determined by NaOH fusion analysis.

The SEP provides several important pieces of evidence for metal natural attenuation. First, analysis of the exchangeable metals in the 1M $MgCl_2$ leach represents the metal source term that is reversibly sorbed. This is also the total metal fraction available for exchange. The exchangeable fraction value more appropriately determines the potentially soluble fraction of metals compared to the DIWET, because metals electrostatically bound to soil colloids are considered bioavailable. The DIWET relies on chemical gradients between the deionized water extracting solution and the soil or sediment matrix. The remaining steps perform two functions: 1) quantify the amount of metal that is irreversibly sorbed, and 2) quantify the amount of metal that is precipitated or coprecipitated as an insoluble metal complexes.

Sequential extraction procedure results for copper, chromium, and zinc in soil with a pH greater than 6 are summarized in Figure 2. Results indicate that the majority of copper is present in the sulfide, carbonate, and hydroxide metal sinks. Metals that are irreversibly sorbed, precipitated, or coprecipitated in solid phases are stable and very insoluble under normal soil conditions. Residence time effects (the longer the metal is in contact with the mineral surface) is a key factor in the long-term stability of these complexes. As shown in figure 2, the majority of metals are present in non-bioavailable forms.

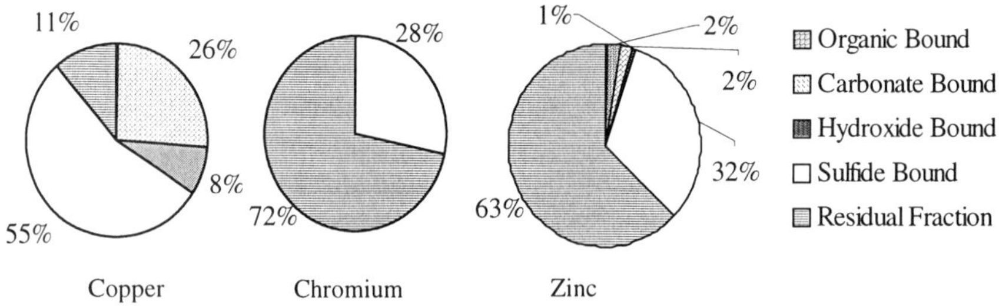

FIGURE 2. Sequential Extraction Results for Copper, Chromium and Zinc.

Chromium was only detected in the sulfide and residual fractions, which are both immobilization sinks. Chromium was not detectable in the exchangeable fraction indicating that very little chromium, if any, is leachable. As shown on Figure 2, zinc was detected in all six of the metal sink fractions. SEP results indicate that the major metal sinks for zinc are the residual, sulfide, oxide, and carbonate fractions with a very small portion present in the exchangeable fraction.

SITE #2 – COPPER, CADMIUM, LEAD, NICKEL, and ZINC

Site #2 contains sediments within an active canal system that were impacted with nickel, cadmium, zinc, and lead as a result of previous practices at an up-gradient facility. *MNAtoolbox* was used to determine likely attenuation pathways for these metals. The major attenuation mechanisms identified for site metals was the sorption and precipitation. The regulators overseeing the site recommended simultaneously extracted metal and acid volatile sulfide analysis (SEM/AVS) to demonstrate metal attenuation and characterize sediment toxicity.

Total Metals and SEM/AVS Analysis. Sediment samples were collected from various locations within the canal system and submitted for total metals and SEM/AVS analysis. The AVS fraction was measured to determine the concentration of reduced sulfide in sediment that has the potential to combine with divalent metals to form insoluble salts, which tend not to be bioavailable. The SEM is a measure of those metals that are available for binding with sulfides. If AVS is greater than SEM, the metals are expected to be bound by sulfides viewed as unavailable biologically (EPA, 1998). In simple terms, SEM/AVS can be used to determine the fraction of metal that is bound to sulfides and therefore not viewed as bioavailable to organisms.

For all samples submitted for SEM/AVS analysis, the concentration of simultaneously extracted metals was less that the acid volatile sulfide fraction indicating that metal toxicity was not expected from sediments. These results provide partial evidence of metal immobilization and reduced bioavailability. Other potential metal sinks (i.e., sorption, and complexation by other soil constituents) are not addressed by the SEM/AVS method.

Although AVS/SEM is assumed to measure the bioavailability of metals associated with sulfides, the acidity of the extracting solution may result in metal solubilization from other sinks. Copper is known to form sparingly soluble oxides and hydroxides at pH 5 and above (McBride, 1994), form carbonate complexes, and is also known to be immobilized by organic matter. These sinks are potentially extracted but not accounted for in an AVS/SEM analysis. Though the latter mechanisms may provide an exposure route to benthic organisms in surface sediments (i.e., attenuation other than sulfide precipitation), their effect on immobilization remains an important attenuation mechanism for deeper sediments below the active benthic layer. In these deeper sediments, immobilization by attenuation mechanisms other than metal sulfide complexation may be effective in preventing metal migration to potential receptors. The method assumes that sulfides provide a large portion of metal sinks (approximately 95%). This method may not be as feasible in sediments collected from areas with characteristically low sulfides or soils where other immobilization processes are more pronounced, as documented by Pardue et al (1998).

DISCUSSION

Natural attenuation of metals is a feasible remedial alternative that requires proper characterization to demonstrate that MNA will be protective of human health and the environment. MNAtoolbox provides an effective tool for screening sites with limited data for potential MNA applications. In addition, other site data (i.e., total and soluble metal analysis, geochemical information, TCLP analysis, etc.) can be used to assist in a preliminary evaluation of the potential for metal MNA at a site.

The data required for evaluating attenuation mechanism may vary with regulating agency and media. Both SEPs and AVS/SEM provide viable methods for characterizing attenuation and bioavailability. SEM/AVS does not specifically identify the attenuation mechanism(s) and may over or under estimate the attenuation capacity and/or bioavailability of metals. This method is also only applicable in areas where sulfide is expected to be in high concentrations (i.e., highly reduced sediments or soils). The SEM/AVS method may be applicable for sites where metal sulfide complexation is viewed as the only metal attenuation mechanism that satisfies both human health and ecological exposure criteria.

The SEPs successfully identified the metal immobilization mechanisms working in soils and provided a mass balance for each of the attenuation sinks. The extraction method used to determine each fraction in the SEP also provides a method for assessing the long-term fate of the immobilized metal. The extractions become more destructive during the SEP analysis, which indicates the relative insolubility of the metal complexes. In addition these extractions provide a worst case scenario by which immobilized metals could be remobilized under extreme conditions.

REFERENCES

Brady P.V., M.V. Brady, and D.J. Borns.1997. *Natural Attenuation: CERCLA, RBCA's and the Future of Environmental Remediation.* Lewis Publisher. New York

Brady P.V., B. P. Spalding, K. M. Krupka, R. D. Waters, P. Zhang, D. J. Borns, and W.D. Brady .1999. *Site Screening and Technical Guidance for Monitored Natural Attenuation at DOE Sites.* Sandia Report (SAND 99-0464)

Chapelle F. H. 1996. "Identifying redox conditions that favor the natural attenuation of chlorinated ethenes in contaminated ground-water systems."*Second Annual International Symposium on Intrinsic Bioremediation: Natural Attenuation.*

McBride, M.B. 1994. *Environmental Chemistry of Soils.* Oxford, New York

Pardue, J.H., and W.A. Jackson. 1998. "Assessment of metal inhibition of reductive dechlorination of hexachlorobenzene at a superfund site"*EnvironmentalToxicolology and Chemistry*. Vol. 17 No 8. pp 1441-1446.

Rice D. W., Dooher B. P., Cullen S. J., Everett L. G., Kastenberg W. E., Grose R. D., and Marino M. A. 1995. *Recommendations to improve the cleanup process for California's leaking underground fuel tanks*. Lawrence Livermore National Laboratory Report UCRL-AR-121762.

U. S. Environmental Protection Agency, 1998, *Contaminated Sediment Management Strategy*, EPA-823-R-98-001, Office of Water, EPA, Washington, D.C., 133 p.

NATURAL ATTENUATION OF 1,2-DICHLOROETHANE IN GROUNDWATER AT A CHEMICAL MANUFACTURING FACILITY

Evan E. Cox and David Major (GeoSyntec Consultants, Guelph, ON, Canada)
Elizabeth Edwards (University of Toronto, Toronto, ON, Canada)

ABSTRACT: 1,2-Dichloroethane (1,2-DCA; also known as ethylene dichloride or EDC) is intrinsically biodegrading through a combination of anaerobic and aerobic processes in the groundwater at a former landfill and ammonium nitrate impoundment at a chemical manufacturing facility in Louisiana. In the source area, 1,2-DCA is undergoing extensive reductive dihaloelimination to ethene, with ethene concentrations exceeding 48 mg/L. Laboratory microcosm studies using site soil and groundwater confirmed the dihaloelimination reaction, with ethene recoveries typically representing between 62 and 99% of the 1,2-DCA mass loss. Intrinsic dihaloelimination half-lives ranged between 33 and 230 days; the addition of acetate as an exogenous electron donor reduced the dihaloelimination half-life to as low as 3 days. Dissolved carbon dioxide (CO_2) concentrations are substantially elevated in source area wells compared to background, suggesting that 1,2-DCA is undergoing oxidation to CO_2. Laboratory microcosm studies confirmed that 1,2-DCA can be oxidized to CO_2 under both anaerobic nitrate-reducing conditions and aerobic conditions, at 1,2-DCA concentrations in excess of 10% of 1,2-DCA solubility. Biodegradation rates for the oxidation reactions were extremely fast, with half-lives ranging between 1 and 8 days. Based on the limited distribution of 1,2-DCA at this site and its documented biodegradability, this site is a strong candidate for natural attenuation as a stand-alone remedy.

INTRODUCTION

1,2-DCA has been extensively used as a feedstock in the production of polyvinyl chloride plastics and a variety of chemical and pharmaceutical products. It is present in the environment as a result of its handling, use and disposal, and currently affects the groundwater of a significant number of industrial facilities in North America and worldwide. The annual production of 1,2-DCA was estimated at 8.2 billion kilograms in 1993, making it the most produced chlorinated solvent (U.S. EPA, 1999). By comparison, the estimated annual production of trichloroethene (TCE) in 1991 was only 0.2 billion kilograms. According to the Toxics Release Inventory, releases of 1,2-DCA to water and land totaled more than 205,000 kilograms between 1987 to 1993. Despite its high production numbers, surprisingly little is known about 1,2-DCA's fate and behavior in groundwater environments compared to its chlorinated solvent counterparts TCE and 1,1,1-trichlororethane (1,1,1-TCA).

Laboratory and field research is beginning to show that 1,2-DCA is susceptible to both aerobic and anaerobic biodegradation. In anaerobic groundwater environments, evidence from most field sites shows that 1,2-DCA preferentially undergoes anaerobic dihaloelimination to ethene (Bosma et al., 1997; Cox et al., 1998; Klecka et al., 1998). The preference for the dihaloelimination reaction results from the more favorable thermodynamics of the reaction versus sequential dechlorination via chloroethane (CA) to ethane. In aerobic environments, direct oxidation of 1,2-DCA to CO_2 has long been known (Janssen et al., 1985), with chloroethanol and chloroacetic acid produced as degradation intermediates (Janssen et al., 1994). Recently, Gerritse et al. (1999) documented the oxidation of 1,2-DCA to CO_2 under nitrate-reducing conditions, a process that is confirmed by data presented herein. Reaction rates for this anaerobic oxidation reaction were very fast, ranging as high as 925 μM of 1,2-DCA transformed per day. Given the recent documentation of this degradation mechanism, little is known about its occurrence and rate in field environments. 1,2-DCA biodegradation reaction pathways are shown in Figure 1.

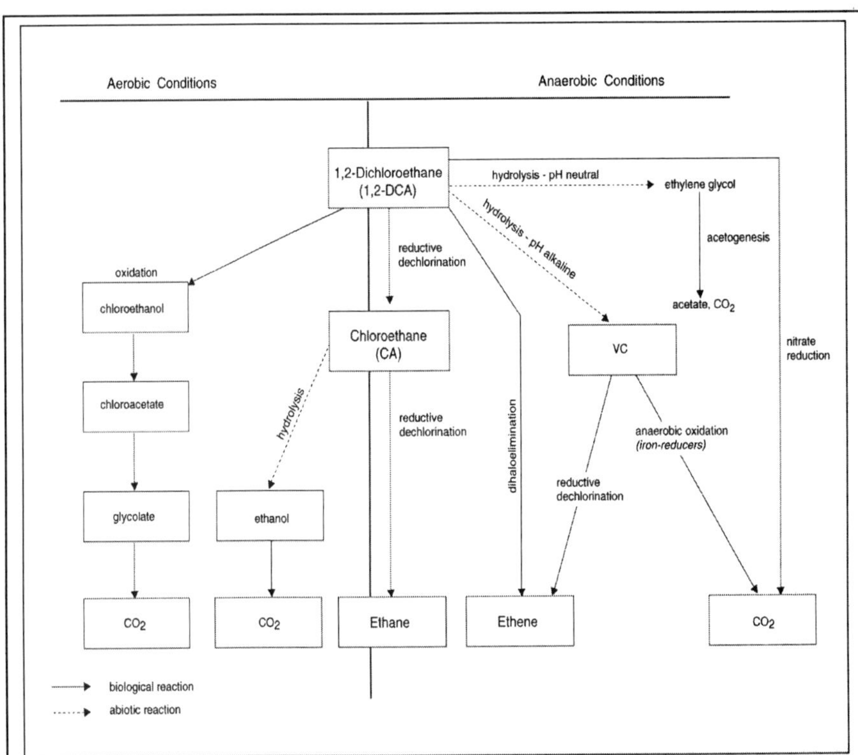

FIGURE 1. Pathways for the Degradation of 1,2-Dichloroethane.

Although 1,2-DCA biodegradation in laboratory studies is now well documented, information on 1,2-DCA biodegradation and bioremediation at field sites does not fully reflect its position as the most produced, and possibly released, chlorinated solvent. In fact, few studies have documented natural attenuation or accelerated bioremediation of 1,2-DCA at field sites, despite the apparent biodegradability of this compound. The present investigation was conducted to examine the biodegradation processes that appear to be controlling the distribution and migration of 1,2-DCA at a chemical manufacturing facility in Louisiana, and to assess and better understand the factors that promote and influence 1,2-DCA biodegradation and bioremediation at high concentrations.

Site Description. The site consists of a shallow 1,2-DCA-impacted aquifer at a former landfill and ammonium nitrate impoundment at a chemical manufacturing facility in Louisiana. The impacted aquifer extends to a depth of approximately 8 m. The depth to the water table fluctuates during the year, from 1 to 4 m below ground surface, reflecting the impact of recharge on this shallow aquifer system. Descending from ground surface, the site geology consists of clayey silt to 3 m, underlain by approximately 5 m of silt and silty sand. The hydraulic conductivity of the aquifer averages 6.7×10^{-5} cm/s. Hydraulic gradients at the site are small, averaging 0.015 m/m. Groundwater velocities at the site are slow, averaging slightly less than 1 m/year. 1,2-DCA appears to be the only chlorinated solvent released at the site, and is the dominant source of carbon in the site groundwater

MATERIALS AND METHODS

The investigation consisted of two components, which were: i) collection of groundwater samples from existing groundwater monitoring wells screened in the impacted aquifer to assess the nature, rate and extent of intrinsic biodegradation processes occurring in situ; and ii) laboratory microcosm studies conducted using site soil and groundwater to confirm degradation processes and assess factors influencing the biodegradation reactions.

Collection and Analysis of Groundwater Samples. Groundwater samples were collected from several groundwater monitoring wells at the site for analysis of 1,2-DCA and suspected degradation intermediates and end products, as well as associated geochemical parameters (redox, potential electron donors and acceptors). Samples were collected following standard sampling and handling procedures, and were analyzed by commercial laboratories following EPA and published methods. Redox parameters (dissolved oxygen and oxidation-reduction potential) were measured during well purging using standard methods.

Collection of Soil and Groundwater for Microcosm Studies. 1,2-DCA-impacted aquifer materials and groundwater were collected from a sample site (#5) located approximately 85 m downgradient from the source area site. Materials were aseptically collected into sterile core liners (3 cm diameter) advanced into the aquifer. Groundwater for the microcosm studies was collected from an existing groundwater monitoring well at location #5. Aquifer cores and groundwater samples were express-shipped on ice to the University of Toronto for microcosm construction.

Microcosm Construction, Incubation and Analysis. Anaerobic microcosms were constructed by aseptically transferring 30 g of homogenized aquifer material (wet weight) and 60 mL of site groundwater into sterile 100-ml (nominal volume) serum bottles. Soil manipulation and microcosm construction for the anaerobic microcosms was conducted in an anaerobic glove box. Selected anaerobic microcosms were amended with 50 mg/L of either acetate or nitrate (both present in the site groundwater) to evaluate the influence of exogenous electron donors or electron acceptors on the degradation rate and extent. Aerobic microcosms were constructed by aseptically transferring 60 g of homogenized aquifer material (wet weight) and 60 mL of site groundwater into sterile 250-ml (nominal volume) serum bottles, allowing a sufficient headspace for oxygen supply. Sterile anaerobic and aerobic control microcosms were autoclaved for 60 minutes at 121°C on three consecutive days and poisoned with mercuric chloride and/or sodium azide. Microcosm bottles were sealed with mininert™ caps to allow repetitive sampling and/or re-spiking. All controls and treatments were incubated in the dark, at room temperature (20°C; similar to groundwater temperature at the site), in an inverted position to minimize VOC loss via the mininert™ cap. Microcosms were periodically sampled for analysis of 1,2-DCA and its suspected degradation intermediates (CA, vinyl chloride [VC], chloroethanol and chloroacetate) and end products (ethene, ethane, and CO_2). Samples were also analyzed periodically for added electron donors or acceptors to determine the need for amendment. Analyses were conducted following accepted published methods.

RESULTS AND DISCUSSION

Characterization of Groundwater Chemistry. Groundwater chemistry data for key monitoring wells on the groundwater flowpath are summarized in Table 1. Redox conditions are primarily anaerobic and reducing in the source area, with dissolved oxygen concentrations of less than 0.5 mg/L and oxidation-reduction potential values ranging between -41 and -147 mV. 1,2-DCA concentrations in the source area range as high as 4,500 mg/L and have remained stable over time, suggesting the presence of some amount of residual 1,2-DCA material. However, 1,2-DCA concentrations decline rapidly with distance from the source area. Ethene concentrations range as high as 48 mg/L within the anaerobic source area, indicating that reductive dihaloelimination to ethene is a primary attenuation mechanism. CA and ethane have only been detected at trace concentrations in groundwater, suggesting that sequential reductive dechlorination is not a significant process at this site. The preferential production of ethene is consistent with observations from other sites (Klecka et al., 1998; Bosma et al., 1998; Cox et al., 1998; Gerritse et al, 1999).

Dissolved CO_2 concentrations are also substantially elevated (up to 437 mg/L) in source area wells compared to background (239 mg/L), suggesting that 1,2-DCA is undergoing oxidation to CO_2. Nitrate and ammonia concentrations are high (up to 110 and 18 mg/L, respectively) in several wells at the site, attributable to impacts the former ammonium nitrate ponds. However, nitrate is typically depleted (<0.05 mg/L) within 1,2-DCA-impacted areas, suggesting that anaerobic oxidation of 1,2-DCA under nitrate-reducing conditions may be an important attenuation mechanism at this site. Chloride concentrations in source area and downgradient wells are an order of magnitude higher than background, providing evidence in support of mineralization of 1,2-DCA.

The site data also provide evidence in support iron- and manganese-reduction, sulfate-reduction, acetogenesis and methanogenesis. In contrast to what we have observed at another 1,2-DCA-impacted site (Cox et al., 1998), there appears to be an inverse correlation between ethene and methane concentrations, with higher ethene concentrations observed at wells having lower methane concentrations (<4 mg/L). By comparison, methane concentrations range as high as 20 mg/L at locations where ethene is absent. Elevated 1,2-DCA concentrations at this site may suppress methanogenesis without limiting the dihaloelimination reaction.

TABLE 1. Groundwater Chemistry Data, Existing Wells

Monitoring Well:	1	2	3	4	5
Distance from Source (m):	Background	0	30	70	85
<u>Redox</u>					
Dissolved Oxygen (mg/L)	0.15	0.27	0.20	0.22	0.28
ORP (mV)	64	-61	-41	-115	-76
<u>Chlorinated Volatiles (mg/L)</u>					
1,2-Dichloroethane	<0.001	4,500	<0.1	700	0.005
Chloroethane	<0.001	<100	<0.1	<25	<0.001
Vinyl Chloride	<0.001	<100	0.48	<25	<0.001
1,2-Dichloroethene (total)	<0.001	<100	<0.1	<25	<0.001
<u>Dissolved Gases (mg/L)</u>					
Ethene	<0.001	48	17	13	0.001
Ethane	<0.001	<0.001	0.003	0.001	<0.001
Methane	4.63	1.29	1.39	3.73	19.8
Carbon Dioxide	239	426	437	388	181
<u>Biochemical Parameters (mg/L)</u>					
Chloride	250	1,200	1,000	780	970
Acetate	<1	<1	<1	8.5	<1
Nitrate	71	0.88	<0.05	<0.05	<0.05
Nitrite	<0.05	<0.05	<0.05	<0.05	<0.05
Ammonia	11	12	3.6	7.2	1.6
Dissolved Iron	0.4	28	5.8	0.51	0.26
Dissolved Manganese	1.4	6.1	7.6	2.1	0.48
Sulfate	90	45	130	66	<5
Sulfide	<0.01	0.23	0.15	0.20	0.13
Biological Oxygen Demand	<2	12	<2	<2	<2
Chemical Oxygen Demand	53	1,400	27	120	63
Dissolved Organic Carbon	22	30	3.9	12	13

Reductive Dihaloelimination of 1,2-DCA to Ethene in Microcosms. Dihaloelimination of 1,2-DCA to ethene was observed in both anaerobic intrinsic controls (no added electron donor) and in microcosms containing acetate. Intrinsic dihaloelimination half-lives ranged between 33 and 230 days; the addition of acetate reduced the dihaloelimination half-life to as low as 3 days. Figure 2 shows the dihaloelimination of 1,2-DCA (10 mg/L) to ethene in the presence of acetate. Ethene recovery in the microcosms ranged between 62 and 99% of the 1,2-DCA mass loss. Re-spikes of the microcosms with progressively higher concentrations of 1,2-DCA showed active dihaloelimination at concentrations exceeding 150 mg/L (Figure 2). However, the rates of dihaloelimination were slower at the higher 1,2-DCA concentrations; half-lives of 35 days were estimated for 1,2-DCA starting concentrations of 75 and 150 mg/L, as compared to 3 to 5 days for 1,2-DCA starting concentrations of 10 and 50 mg/L. Consistent with field data, CA and ethane were not detected in either control or treatment microcosms. Methane concentrations remained relatively stable over the incubation period, showing little to no correlation with dihaloelimination (data not presented).

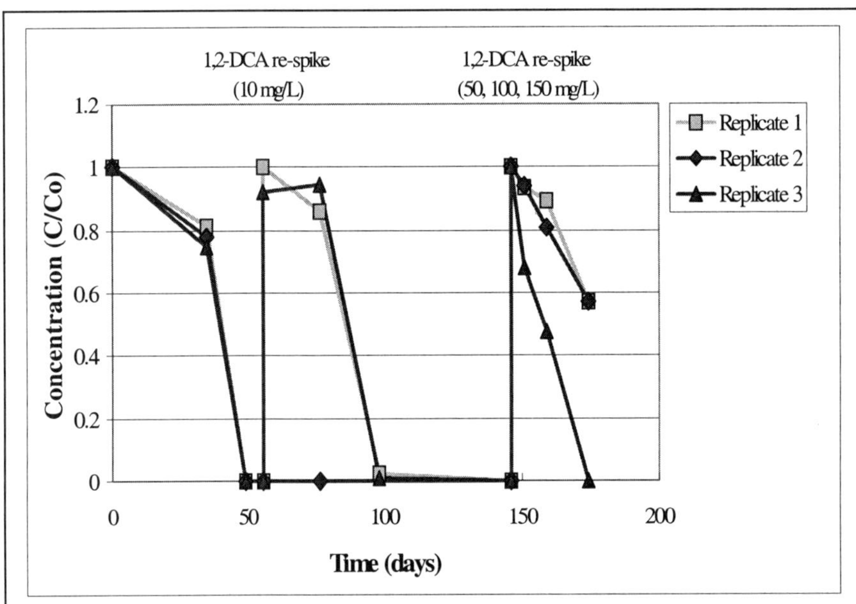

FIGURE 2. Reductive dihaloelimination of 1,2-DCA in the presence of acetate. Initial 1,2-DCA concentration of 10 mg/L. First re-spike = 10 mg/L. Second re-spike = 150 mg/L (replicate 1), 100 mg/L (replicate 2) and 50 mg/L (replicate 3).

Anaerobic Oxidation of 1,2-DCA under Nitrate-Reducing Conditions. Biodegradation of 1,2-DCA in the presence of available nitrate was very rapid (Figure 3), with little apparent effect of 1,2-DCA concentration on degradation rate. Anaerobic oxidation half-lives consistently ranged between 1 and 6 days, over starting 1,2-DCA concentrations of 10 to 1,000 mg/L (>10% of 1,2-DCA solubility). While historic impacts of the former ammonium nitrate ponds are not documented, they may explain the ready acclimation of microbial populations to anaerobically oxidize high concentrations of 1,2-DCA when nitrate is present.

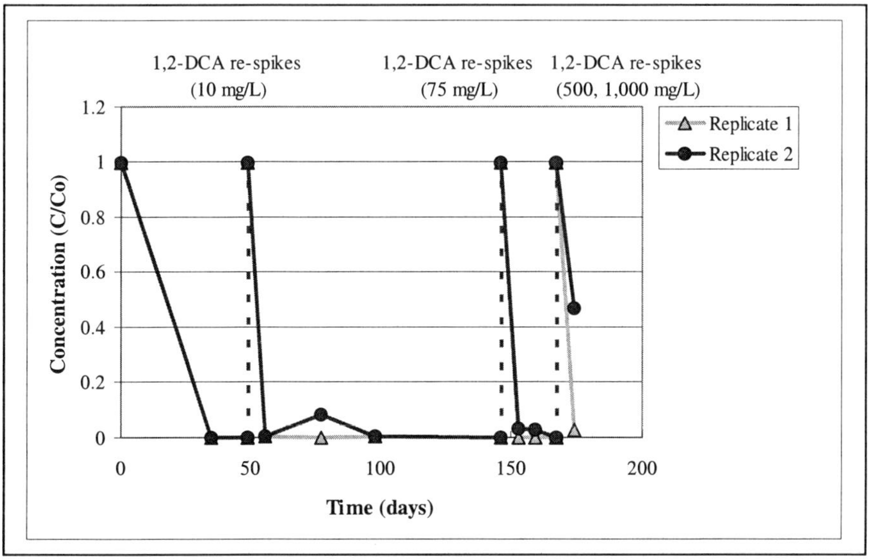

FIGURE 3. Anaerobic oxidation of 1,2-DCA under nitrate-reducing conditions. Initial 1,2-DCA concentration of 10 mg/L. First re-spike = 10 mg/L. Second re-spike = 75 mg/L. Third re-spike = 500 mg/L (rep 1) and 1,000 mg/L (rep 2).

Aerobic Oxidation of 1,2-DCA. Rapid and complete aerobic oxidation of 1,2-DCA was observed in the aerobic microcosms, at starting concentrations ranging from 10 to 1,000 mg/L. As with the nitrate-reducing microcosms, little apparent effect of 1,2-DCA concentration on degradation rate was observed. 1,2-DCA oxidation half-lives ranged from 2 to 5 days at starting 1,2-DCA concentrations of 10 mg/L; and 8 days at starting 1,2-DCA concentrations in excess of 800 mg/L. Chloroethanol and chloroacetate were transiently observed in the microcosms; CO_2 was observed as the final product.

CONCLUSIONS

The results of these investigations provide evidence of the mechanisms that are controlling the migration and distribution of 1,2-DCA in the site groundwater. Within the anaerobic source area, 1,2-DCA is undergoing significant reductive dihaloelimination to ethene. Anaerobic oxidation to CO_2 under nitrate-reducing conditions also appears to be a significant process based on microcosm results, although its contribution under field conditions is difficult to quantify based solely on data from monitoring wells. Aerobic oxidation is also likely to contribute significantly to the observed 1,2-DCA attenuation in areas of the plume (e.g., the lateral and vertical fringes) where re-aeration occurs due to recharge. The field data indicate that reductive dihaloelimination of 1,2-DCA to ethene occurs within the source area where 1,2-DCA concentrations exceed 50% of its solubility. Similarly, the microcosm data show that anaerobic and aerobic oxidation processes are capable of biodegrading 1,2-DCA at concentrations in excess of 10% of its solubility. The implications of these data for site remediation are important because they show that natural attenuation and accelerated bioremediation can be acceptable for treatment/control of source areas in addition to dissolved plumes. For this site, the combination of slow groundwater velocity, high intrinsic biodegradation rates, and the absence of unacceptable risks appears to make monitored natural attenuation an appropriate remedy for the site.

REFERENCES

Bosma T.N.P., M.A.van Aalast, H.H.M. Rijnaarts, J. Taat, J. Bovendeur. 1997. "Intrinsic dechlorination of 1,2-dichloroethane at an industrial site." In B.C. Alleman and A. Leeson (Eds.), *In Situ and On-Site Bioremediation: Volume 3*. Battelle Press, Columbus, OH.

Cox, E.E., M. McMaster and D.W. Major. 1998. "Natural attenuation of 1,2-dicloroethane and chloroform in groundwater at a Superfund site." In G.B. Wickramanayake and R.E. Hinchee (Eds.), *Remediation of Chlorinated and Recalcitrant Compounds*: *Volume 1*. Battelle Press, Columbus, OH.

Gerritse, J., A.Borger, E. van Heiningen, H.H.M. Rijnaarts, T. Bosma, J. Taat, B. van Winden, J. Dijk, and J.A.M. de Bont. 1999. "Assessment and monitoring of 1,2-Dichlorethane dechlorination." In A. Leeson and B.C. Alleman (Eds.), *In Situ and On-Site Bioremediation: Volume 5(2)*. Battelle Press, Columbus, OH.

Janssen D.B., J.R. van der Ploeg, and F. Pies. 1994. "Genetics and biochemistry of 1,2-dichloroethane degradation". Biodegradation. 5:249-257

Janssen, D.B., A. Scheper, L. Dijkhuizen, and B. Witholt. 1985. "Degradation of halogenated aliphatic compounds by *Xanthobacter autotrophicus* GJ10". Appl. Environ. Microbiol. 49:673-677.

Klecka, G. M., C.L. Carpenter, and S.J. Gonsier. 1998. "Biological transformations of 1,2-dichloroethane in subsurface soils and groundwater." J. Contam. Hydrology. 34: 139-154.

2000 AUTHOR INDEX

This index contains names, affiliations, and book/page citations for all authors who contributed to the seven books published in connection with the Second International Conference on Remediation of Chlorinated and Recalcitrant Compounds, held in Monterey, California, in May 2000. Ordering information is provided on the back cover of this book.

The citations reference the seven books as follows:

2(1): Wickramanayake, G.B., A.R. Gavaskar, M.E. Kelley, and K.W. Nehring (Eds.), *Risk, Regulatory, and Monitoring Considerations: Remediation of Chlorinated and Recalcitrant Compounds*. Battelle Press, Columbus, OH, 2000. 438 pp.

2(2): Wickramanayake, G.B., A.R. Gavaskar, and N. Gupta (Eds.), *Treating Dense Nonaqueous-Phase Liquids (DNAPLs): Remediation of Chlorinated and Recalcitrant Compounds*. Battelle Press, Columbus, OH, 2000. 256 pp.

2(3): Wickramanayake, G.B., A.R. Gavaskar, and M.E. Kelley (Eds.), *Natural Attenuation Considerations and Case Studies: Remediation of Chlorinated and Recalcitrant Compounds*. Battelle Press, Columbus, OH, 2000. 254 pp.

2(4): Wickramanayake, G.B., A.R. Gavaskar, B.C.Alleman, and V.S. Magar (Eds.) *Bioremediation and Phytoremediation of Chlorinated and Recalcitrant Compounds*. Battelle Press, Columbus, OH, 2000. 538 pp.

2(5): Wickramanayake, G.B. and A.R. Gavaskar (Eds.), *Physical and Thermal Technologies: Remediation of Chlorinated and Recalcitrant Compounds*. Battelle Press, Columbus, OH, 2000. 344 pp.

2(6): Wickramanayake, G.B., A.R. Gavaskar, and A.S.C. Chen (Eds.), *Chemical Oxidation and Reactive Barriers: Remediation of Chlorinated and Recalcitrant Compounds*. Battelle Press, Columbus, OH, 2000. 470 pp.

2(7): Wickramanayake, G.B., A.R. Gavaskar, J.T. Gibbs, and J.L. Means (Eds.), *Case Studies in the Remediation of Chlorinated and Recalcitrant Compounds*. Battelle Press, Columbus, OH, 2000. 430 pp.

2000 KEYWORD INDEX

This index contains keyword terms assigned to the articles in the seven books published in connection with the Second International Conference on Remediation of Chlorinated and Recalcitrant Compounds, held in Monterey, California, in May 2000. Ordering information is provided on the back cover of this book.

In assigning the terms that appear in this index, no attempt was made to reference all subjects addressed. Instead, terms were assigned to each article to reflect the primary topics covered by that article. Authors' suggestions were taken into consideration and expanded or revised as necessary to produce a cohesive topic listing. The citations reference the seven books as follows:

2(1): Wickramanayake, G.B., A.R. Gavaskar, M.E. Kelley, and K.W. Nehring (Eds.), *Risk, Regulatory, and Monitoring Considerations: Remediation of Chlorinated and Recalcitrant Compounds*. Battelle Press, Columbus, OH, 2000. 438 pp.

2(2): Wickramanayake, G.B., A.R. Gavaskar, and N. Gupta (Eds.), *Treating Dense Nonaqueous-Phase Liquids (DNAPLs): Remediation of Chlorinated and Recalcitrant Compounds*. Battelle Press, Columbus, OH, 2000. 256 pp.

2(3): Wickramanayake, G.B., A.R. Gavaskar, and M.E. Kelley (Eds.), *Natural Attenuation Considerations and Case Studies: Remediation of Chlorinated and Recalcitrant Compounds*. Battelle Press, Columbus, OH, 2000. 254 pp.

2(4): Wickramanayake, G.B., A.R. Gavaskar, B.C.Alleman, and V.S. Magar (Eds.) *Bioremediation and Phytoremediation of Chlorinated and Recalcitrant Compounds*. Battelle Press, Columbus, OH, 2000. 538 pp.

2(5): Wickramanayake, G.B. and A.R. Gavaskar (Eds.), *Physical and Thermal Technologies: Remediation of Chlorinated and Recalcitrant Compounds*. Battelle Press, Columbus, OH, 2000. 344 pp.

2(6): Wickramanayake, G.B., A.R. Gavaskar, and A.S.C. Chen (Eds.), *Chemical Oxidation and Reactive Barriers: Remediation of Chlorinated and Recalcitrant Compounds*. Battelle Press, Columbus, OH, 2000. 470 pp.

2(7): Wickramanayake, G.B., A.R. Gavaskar, J.T. Gibbs, and J.L. Means (Eds.), *Case Studies in the Remediation of Chlorinated and Recalcitrant Compounds*. Battelle Press, Columbus, OH, 2000. 430 pp.

A

D

E

F

M

N

R

S

T

U–Z